THE PURSUIT OF NATURE

INFORMAL ESSAYS ON
THE HISTORY OF
PHYSIOLGOY

THE PURSUIT OF NATURE

INFORMAL ESSAYS ON THE HISTORY OF PHYSIOLOGY

A. L. HODGKIN, A. F. HUXLEY, W. FELDBERG
W. A. H. RUSHTON, R. A. GREGORY, R. A. McCANCE

These essays were written as part of the celebrations of
the Centenary of the Physiological Society in 1976

CAMBRIDGE UNIVERSITY PRESS

CAMBRIDGE

LONDON · NEW YORK · MELBOURNE

CAMBRIDGE UNIVERSITY PRESS
Cambridge, New York, Melbourne, Madrid, Cape Town, Singapore,
São Paulo, Delhi, Dubai, Tokyo, Mexico City

Cambridge University Press
The Edinburgh Building, Cambridge CB2 8RU, UK

Published in the United States of America by Cambridge University Press, New York

www.cambridge.org
Information on this title: www.cambridge.org/9780521296175

First published 1977
First paperback edition 1979
Re-issued 2010

A catalogue record for this publication is available from the British Library

Library of Congress Cataloguing in Publication data
Main entry under title:
The Pursuit of Nature.
'These essays were written as part of the celebrations of the centenary of
the Physiological Society in 1976.'
Includes index.
1. Physiology – History – Addresses, essays, lectures.
I. Hodgkin, Alan Lloyd. II. Physiological Society.
QP21. P87   591.1'09   76–58844

ISBN 978-0-521-21505-3 Hardback
ISBN 978-0-521-29617-5 Paperback

CONTENTS

LIST OF CONTRIBUTORS

SIR ALAN L. HODGKIN	Physiological Laboratory, Downing Street, Cambridge CB2 3EG
SIR ANDREW F. HUXLEY	Department of Physiology, University College, Gower Street, London WC1E 6BT
PROFESSOR W. FELDBERG	National Institute for Medical Research, Mill Hill, London NW7 1AA
PROFESSOR W. A. H. RUSHTON	Trinity College, Cambridge CB2 1TQ
PROFESSOR R. A. GREGORY	Physiological Laboratory, University of Liverpool, PO Box 147, Liverpool L69 3BX
PROFESSOR R. A. MCCANCE	Sidney Sussex College, Cambridge CB2 3HU

CHANCE AND DESIGN IN ELECTROPHYSIOLOGY: AN INFORMAL ACCOUNT OF CERTAIN EXPERIMENTS ON NERVE CARRIED OUT BETWEEN 1934 AND 1952

By A. L. HODGKIN

My aim in this lecture is to give you some idea of the informal background to the series of papers on nerve conduction which my colleagues and I wrote between the years 1937 and 1952. The sort of questions which I wish to consider are these: when was the work started? And why? How much was done by accident and how much by careful planning? Was the equipment found to be satisfactory in its original form or did it evolve gradually? What books, papers or people determined the choice of that particular piece of research? And so on. Such recollections are likely to be somewhat personal and may not be fair to others. Yet I think it necessary to give such an account, because I believe that the record of published papers conveys an impression of directness and planning which does not at all coincide with the actual sequence of events. The stated object of a piece of research often agrees more closely with the reason for continuing or finishing the work than it does with the idea which led to the original experiments. In writing papers, authors are encouraged to be logical, and, even if they wished to admit that some experiment which turned out in a logical way was done for a perfectly dotty reason, they would not be encouraged to 'clutter-up' the literature with irrelevant personal reminiscences. But over a long period I have developed a feeling of guilt about suppressing the part which chance and good fortune played in what now seems to be a rather logical development.

I can illustrate some of these points by considering the 'history' (if that is not too grand a word) of my first two papers, which were published in the *Journal of Physiology* in 1937 under the title 'Evidence for electrical transmission in nerve'. The aim of the papers is stated in the first two sentences: the method is straightforward; fire an impulse at a local block and see what happens beyond – and the result now seems so obvious that one wonders whether the work was worth doing at all. I suspect that this was one of the papers which caused a very distinguished biologist to say, 'The trouble with you Cambridge electrophysiologists is that you never discover anything; you think hard, decide what is right and then work

away until you prove it'. In defending myself and my colleagues against
this accusation I must come clean and dispel any impression of tidy planning which our papers may have created.

I first started the block experiments in the summer of 1934. I was then
a second-year undergraduate, undecided whether to read Physiology or
Zoology in Part II. My inclination was towards the former but I was
advised that there were no prospects for physiologists without medical
degrees. Curiously enough the thing that finally converted me to Physiology
was a rather unfair remark made by my director of studies, Jack Roughton,
who said, 'All experimental zoologists do is apply to many animals the
conclusions which physiologists have reached by working on one particular
animal; if you want to find out anything really new you must join us in
Physiology'. As the Physiological Laboratory in Cambridge then contained
J. Barcroft, E. D. Adrian, B. H. C. Matthews, Grey Walter, F. J. W.
Roughton, G. S. Adair, E. N. Willmer, F. R. Winton and several other
distinguished physiologists, there was something to be said on Roughton's
side. At all events I was converted by the remark although it cannot
be defended against the examples of J. Gray, D. Keilin, H. W. Lissmann,
P. B. Medawar, C. F. A. Pantin, J. W. S. Pringle, V. B. Wigglesworth
and J. Z. Young, to name only a few of the very distinguished British
scientists who have approached biology from the zoological side.

During my first two years at Cambridge I had become interested in
membranes, mainly through reading James Gray's (1931) *Experimental
Cytology* and A. V. Hill's (1932) *Chemical Wave Transmission in Nerve*,
both of which were then relatively new. I had also read the excellent review
by Osterhout (1931) on 'Physiological studies of large plant cells' and was
impressed with the evidence obtained by Blinks (1930) for an increase of
membrane conductivity during the action potential of *Nitella*. It seemed
to me that this crucial piece of evidence was lacking in nerve and I tried
to test it by the method illustrated in Fig. 1 A. Using class apparatus, some
of which had been designed by Keith Lucas[1] twenty years earlier, I
arranged to block a nerve locally by freezing it, and applied two pairs of
shocks to it in the position shown. I argued that if the permeability and
conductivity of the membrane increased during activity, then arrival of an
impulse at the block should increase the fraction of current which penetrated the nerve and hence lower the electrical threshold at the distal pair
of electrodes. I set up the equipment, using a silver rod and a tin of ice and
salt to cool the nerve, a Keith Lucas spring contact-breaker to time the

---

[1] I was much influenced by Keith Lucas's collected papers which I read all
through as a student, partly at least because I knew his widow and sons. My father
and Lucas were close friends; both died in the first world war; Lucas in an aeroplane
accident in 1917 and my father of dysentery in Baghdad in 1918.

two shocks and a smoked drum to measure the size of muscular contraction. The experiment consisted of establishing a block, and then alternating between shock 2 by itself, and shock 2 preceded by shock 1 with an interval of a few milliseconds between them. If there was any facilitating effect then the muscle twitch evoked by the two shocks was greater than that produced by one alone. I first tried the experiment in July 1934. I got a negative result, but on trying again in October the experiment worked and I was very pleased indeed. Later that year I made a note to the effect that

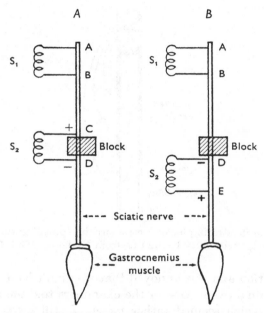

Fig. 1. Diagram of method of testing the effect on excitability of a blocked nerve impulse, using sciatic gastrocnemius preparation.

'Freezing must be light and reversible. In long vac term [i.e. July] non-reversible freezing or ligaturing used and no effect could be detected'. The threshold near the block was very variable and it was difficult to obtain quantitative results. However, a number of controls established the genuine nature of the effect; for example it was abolished by crushing between B and C, was unaffected by reversing A and B, and developed with a shock separation consistent with the time taken for the impulse to travel from B to C.

Then I had a horrid surprise. I switched the anode from just above the block to a position beyond it, i.e. from C to E, as in Fig. 1 *B*, and found that the effect still persisted. It therefore had nothing to do with an increase in membrane conductivity and was most simply explained by

assuming that local circuits were spreading through the block and raising
excitability beyond it, as shown in Fig. 2. More generally, one might
attribute the effect to whatever agent was responsible for conduction. The
existence of the effect did not provide any evidence for electrical transmission, but it offered a nice way of testing the theory. At the time I was
disappointed that I had not obtained any evidence for an increase in
membrane conductivity and I gave up the experiments for the undergraduate Part II work that I should have been doing all along. I specialized

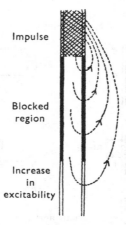

Fig. 2. Diagram illustrating local electric circuits spreading through block
and increasing excitability beyond it (from Hodgkin, 1936, 1937 *a, b*).

on nerve conduction and was annoyed that we didn't have any questions
on nerve in the final exam. One of the examiners told me later that this
was deliberate, which seemed rather mean – I still have an occasional
nightmare about taking exams.

During my Part II year I read all the papers of the St Louis School,
J. Erlanger, H. S. Gasser, G. H. Bishop, H. T. Graham, R. Lorente de Nó
and F. O. Schmitt. This made it clear to me that the leading axonologists
were thoroughly sceptical both of the membrane theory in general and of
the local circuit theory in particular. I came to the conclusion that it
would be well worth while to see whether the transient increase of excitability beyond a localized block was an electrical effect, and decided that I
would take up this as a research project.

In those days laboratory life was rather informal, at any rate in Cambridge. I never worked for a Ph.D. and didn't have a research supervisor.
You might easily start in a bare room and have to build most of your
equipment yourself, apart from a few standard bits like smoked drums,
Palmer stands and kymographs. This sounds depressing but it actually

wasn't. Nowadays, scientists are apt to become neurotic and give up work if they know their equipment is markedly inferior to other people's. Indeed it is regarded as somewhat unscientific to carry out experiments with anything but the best equipment. This certainly wasn't my feeling when I started research and, to begin with, all that mattered was that one should have enough equipment to do something new. I may have been rather extreme in this respect but the general attitude to equipment was certainly very different from that existing today. In his comments on the pre-war Cavendish Laboratory in Cambridge, the distinguished physicist J. A. Ratcliffe (1975) explains that in the 1920s and 30s an elegant piece of apparatus or an elegant experiment meant one that could be built or carried out very cheaply. He says,

There was, I think, a feeling that the best science was that done in the simplest way. In experimental work, as in mathematics, there was 'style' and a result obtained with simple equipment was more elegant than one obtained with complicated apparatus, just as a mathematical proof derived neatly was better than one involving laborious calculations. Rutherford's first disintegration experiment, and Chadwick's discovery of the neutron had a 'style' that is different from that of experiments made with giant accelerators.

Ratcliffe illustrates his point with this anecdote:

...as a young research student I wished to try out a radiocircuit, in the way that was then common, by screwing some components to a wooden 'bread board'. When I went to get a piece of wood for the purpose Lincoln [the head of the workshop] pointed to a pile of scrap wood in the corner and invited me to take a piece, but as I was leaving the room he ran after me and said 'Here, Mr. Ratcliffe, do you really need mahogany?'

I was lucky because I inherited a Matthews oscilloscope and other electrical equipment from Grey Walter. In those days it wasn't considered proper to use an amplifier built by someone else. So I constructed a condenser-coupled triode amplifier in a series of biscuit tins which I painted bright blue. At that time there were no electric soldering irons, no resin-cored solder, and the valves, which were usually microphonic, needed anti-vibration mountings, so building an amplifier took longer than it would today. But I was helped in this and other things by Martin Wright who already showed signs of the mechanical ingenuity and inventiveness for which he is now well known. I also received much help from Charles Morley Fletcher who worked next door to me on *Mytilus* muscle.

In the autumn of 1935 I managed to record the electrotonic potential produced by local electric circuits spreading through the blocked region. However, the Matthews oscilloscope wasn't really fast enough – although admirable for recording the presence of impulses – and the whole set-up was terribly cumbersome, with an arc lamp, rotating mirrors, moving film and a cylindrical paper screen all arranged to give the same effect as a modern cathode ray oscilloscope. So I bought a cathode ray tube and accessories from Cossor, and a second-hand film camera from Wardour Street in London. I also got our workshop mechanic, Mr Hall, to build a rotary contact breaker for starting the sweep and timing two shocks. This did what I wanted but made an incredible din as a series of huge cams smacked into three magneto contact-breakers ten times a second. As I had to pay for this equipment myself I bought the cheapest kind of cathode ray tube; this was a 'soft' tube in which electrons are kept in a column by positively charged gas molecules, rather than by focusing electrodes. Bryan Matthews was away in the Andes that year, but I received much help and advice from A. F. Rawdon-Smith, who had a first-class knowledge of electronics and worked next door in Psychology. In the end all the equipment worked well though I had terrible and quite unnecessary trouble with it. At that time there was a sort of mystical idea that the noisiness of an amplifier varied inversely with the skill of the man who built it, and amplifier noise was regarded as a sort of moral penalty for bad workmanship. As I have always been rotten at making things I naturally attributed my noisy base line to poor workmanship. In fact, as I eventually discovered, the base line in my set-up was relatively noisy because the frequency response of my cathode ray tube was very much higher than that of the Matthews oscilloscope used in the basement. Then the whole business of shock-artifacts was shrouded in mystery and I didn't learn to think rationally on this subject until I went to the Rockefeller Institute in 1937. There I met Dr Toennies, who looked after the electronics in Gasser's group; he told me to forget about radiation fields and other irrelevant ideas that I had been struggling with, and to think only in terms of electrical leaks, stray capacities and actual spread of current in the tissue.

By mid July 1936 I had been through the main experiments and wrote up the results in a fellowship thesis for Trinity College, where I had lived happily for four years. I was surprised and very pleased to be successful but also a little alarmed to be joining a society which included people like J. J. Thomson (the Master), Rutherford, Aston, Eddington, Gowland, Hopkins, Adrian, Wittgenstein, Hardy and Littlewood. After getting a fellowship I spent several months repeating and tidying up experiments; eventually two papers were published in the *Journal of Physiology* (1937) almost exactly three years after the beginning of the experiments. The conclusions were more or less all right except that for a myelinated nerve one would nowadays redraw Fig. 2 to show the current concentrated at the nodes. I think I was lucky not to be completely messed up by the electrical polarizability of the nerve sheath (perineurium) and I have wondered since whether this may not have been made leaky by the ice crystals which formed in the cooled region of nerve.

Chance and good fortune were equally important in my next piece of research. I had grown interested in cable theory and had come to the conclusion that it was necessary to excite a finite length of nerve in order to start an action potential. The argument, which was developed independently and much further by Rushton (1937), led to the idea of a subthreshold response which might explain the unexpected results obtained by Katz (1937) and Rushton (1932) in their studies of excitability. But I didn't make any deliberate attempt to test these ideas and instead followed a suggestion by Professor Adrian that I should work on crab nerve. I am now not sure what I intended to do, but I think I hoped to test the idea that accommodation might be due to the polarization of some structure in series with the nerve fibres. I found it very easy to split crab nerve into fine strands and one of the strands I picked up turned out to be a single axon – to judge from its enormous all-or-nothing action potential. This really was a great piece of luck as I had no dissecting microscope and the chances of picking up one of the half-dozen or so 30 μm fibres in a nerve trunk a millimetre thick are not very high. The next day I borrowed a dissecting microscope and from that time to this I have never worked on a multifibre preparation again. (What, never? Well, hardly ever.)

Soon after I got the preparation going I noticed that a shock which was just below threshold produced something like a small graded action potential, which grew rapidly in size as the stimulus approached threshold. This clearly was exactly what was needed to explain Bernard Katz's results and I was very pleased to be getting evidence of something as unorthodox as a graded response in a single nerve fibre. My electrical technique wasn't really up to recording from single axons as you can see if you look at the illustration in the preliminary note describing the Cambridge experiments (Hodgkin, 1937c). However, help was at hand because Herbert Gasser, who was then Director of the Rockefeller Institute in New York had invited me to spend a year working in his group, and I had been awarded a travelling fellowship by the Rockefeller Foundation. Soon after I arrived, Dr Toennies, the electronics expert in Gasser's group, pointed out that it was essential to use a cathode follower if one wished to make accurate recordings of rapid changes from a high resistance preparation like a single crab axon. He provided the necessary equipment and I learnt a great deal about electronics and electrical recording from Gasser and his group which included Lorente de Nó, Grundfest, Toennies and Hursch.

At that time the Rockefeller Institute was a very distinguished laboratory – as indeed it still is. At lunch time the great men led their flocks to separate tables and one would see little processions headed by Landsteiner, Carrel, Avery, P. A. Levene, van Slyke and so on. It was a pretty formal place and I missed the free and easy casualness of the Cambridge laboratories. But it was a valuable experience to work in a big well-organized laboratory and helped to turn me from an amateur into a professional scientist. Apart from Gasser's own group the people who influenced me most were Osterhout (large plant cells), Michaelis (membranes) and MacInnes and Shedlovsky (electrochemistry), not to mention Peyton Rous and his family on the personal side.

But the contact which had the greatest immediate effect on my scientific life was with Cole and Curtis at Columbia. I was still anxious to know whether the membrane conductance increased during activity and had obtained some evidence that it did, by showing that a shock applied at the crest of the spike produced less than half the normal polarization (Hodgkin, 1938, p. 107). I had also obtained a positive effect in preliminary experiments with alternating currents but the results were untidy because the out-of-balance signal in the bridge was mixed up with a diphasic action potential. I abandoned these rather amateur attempts after I had visited Columbia University and seen the beautiful experiments which Cole and Curtis had already done on *Nitella* and were planning on squid axons (having studied the passive transverse impedance in the previous year). Cole asked me to visit him at Woods Hole in June, and at some point during the spring of 1938 we agreed that I should bring up equipment for measuring the membrane resistance of squid axons by a modification of the resistance-length method used by Rushton (1934) (see Cole & Hodgkin, 1939).

Meanwhile I visited St Louis, on the way to a 3 week holiday in Mexico, which was then a wild and remote country, and also an incredibly inexpensive one where you could live for less than a dollar a day. In St Louis I stayed with Joseph Erlanger who was exceedingly nice to me, but expressed total disbelief in subthreshold activity in myelinated axons and was also very sceptical about the local circuit theory. I had tried hard but without any success to isolate single myelinated axons from cat spinal roots so I knew I could not win on that front. But there was more hope on the other and I got a good idea from a conversation with Erlanger in which he said that he might be convinced if I could alter conduction velocity by changing the electrical resistance outside a nerve fibre. Somewhere on that holiday, which included a four-day train trip from Mexico City to New York, I saw that it would be very easy to alter the electrical resistance outside a crab fibre from a high value in oil to a low one in a large volume of sea water. I did the experiment as soon as I got back to New York and obtained a large effect on conduction velocity (Fig. 3). This was one of the few occasions on which everything went according to plan and this time no hidden snags emerged. I showed Harry Grundfest the records next day and remember that he shook me by the hand like a character in a novel by

C. P. Snow. When I got to Woods Hole in June, Cole and Curtis very kindly let me use their amplifier and I was able to repeat the experiments on squid axons as well as making some other tests of the local circuit theory.

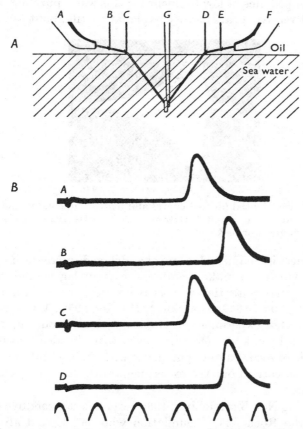

Fig. 3. *A*, arrangement for comparing the conduction velocity of a single nerve fibre in sea water and oil. As shown, the conduction stretch is in a large volume of sea water; on raising the electrode assembly, the volume of conducting fluid is reduced to the thin film of sea water which clings to the fibre in oil.

B, effect of external resistance on conduction velocity. *A* and *C*, action potential recorded with sea water covering 95 % of intermediate conduction distance; *B* and *D*, fibre completely immersed in oil. Conduction distance 13 mm. Time, msec (from Hodgkin, 1939).

This was a very exciting time to be at Woods Hole. I remember arriving in Cole and Curtis's laboratory and seeing the increase in membrane conductance during the action potential displayed in a striking way on the cathode ray tube (Fig. 4). After learning to clean squid axons and repeating the velocity experiments I settled down with Cole to measure resistance length curves. Towards the end of this work Cole noticed that there appeared to be something like an inductance which showed up in the longitudinal impedance at low frequencies; this was a puzzling observation which did not receive a satisfactory explanation till about ten years later

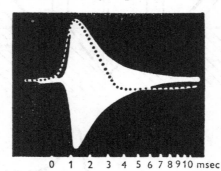

0   1   2   3   4   5 6 7 8 9 10 msec

Fig. 4. Action potential (dotted curve) and increase in conductance (white band) in squid axon at about 6° C (from Cole & Curtis, 1939; see also Cole & Curtis, 1938, and Curtis & Cole, 1938).

(the inductance is mainly due to the delayed increase in potassium conductance which can make membrane current lag behind voltage provided the internal potential is positive to the potassium equilibrium potential (Cole, 1941, 1947; Hodgkin & Huxley, 1952d)). Curtis and I also did a few long-shot experiments trying to push electrodes up the cut end of a giant axon. I think we both came away with the idea that it might not be too difficult to record action potentials with an internal electrode; at all events, we both carried out the experiment with different partners in the following year.

Before leaving New York in July 1938 I went to say goodbye to Dr R. A. Lambert at the Rockefeller Foundation who had looked after me with great kindness during the preceding year. Someone, probably Herbert Gasser, had suggested that the Foundation might provide me with an equipment grant and Toennies had helped to prepare a list of some of the things I might need. I mentioned this to Lambert and was electrified to learn later that I might expect a sum of £300, an unheard of amount for a young scientist in those days.

When I got back to Cambridge in the autumn I decided to set up the kind of equipment used in the Rockefeller, with racks, electronic timing, d.c. amplifiers and so on. I joined forces with A. F. Rawdon-Smith, K. J. W. Craik and R. S. Sturdy in Psychology, and between us we built three or four sets of equipment some of which were still in use twenty-five years later. I did a bit of wiring but the three psychologists did nearly all the work. Rawdon-Smith designed the d.c. amplifier and most of the rest of the equipment, but I had help from Toennies over cathode followers, Otto Schmitt over multivibrators, and Matthews over the camera and many other details. I also remember consulting Britton Chance who was then working in the Physiological Laboratory in the Roughton–Millikan suite.

In the late 30s we were becoming 'professionals' and the objective in designing electronic equipment was not to make some neat miniaturized unit but to build up as massive and imposing an array of racks and panels as you could get – possibly with the idea of cowing your scientific opponents or dissuading your rivals from following in your footsteps. These large units were a nuisance if you wished to move to Plymouth or Woods Hole, but they did have the great advantage of being difficult to borrow when you were on holiday or writing-up results.

It took three or four months to get all the equipment built and to be ready for experiments again. I had worked very hard for the previous six years and as there was obviously going to be a European war I thought it best to choose a straightforward problem which would leave time for non-scientific activities. So I decided to use my new d.c. amplifier to check how close the action potential came to the resting potential. Andrew Huxley, who was doing the Part II Physiology course, joined me in some of the experiments. We measured external electrical changes in *Carcinus* axons immersed in oil and took the resting potential as the steady p.d. between an intact region and one depolarized by injury or isotonic potassium chloride. We found that the action potential was much larger than the resting potential, for example 73 mV for the former as against 37 mV for the latter. Although I wasn't aware of it till much later, Schaefer (1936) had previously reported a similar discrepancy in the sciatic and gastrocnemius muscles of the frog. I got the same result with lobster axons, a preparation which Rushton and I studied later that year in order to calculate passive electrical constants, using the cable-equations which he and others had developed (Hodgkin & Rushton, 1946). These results required much analysis and were put on one side until 1945 when the war was nearly over and my part in it was finished.

The results with external electrodes did not give the absolute value of the action potential and resting potential, because of the short-circuiting effect of the external fluid. But there was no reason why this should affect one potential more than the other, and the difference seemed much too large to be explained by some minor difference in the way the two potentials were recorded.

I decided to continue the experiments at Plymouth where I had worked several times since my first visit there as a schoolboy in 1931. I bought a trailer which I attached to my ancient car and with some difficulty managed to drag the bulk of my equipment from Cambridge to Plymouth in July 1939. After a few weeks, Andrew Huxley joined me and started to measure the viscosity of axoplasm by seeing how fast a mercury droplet would fall down the axis cylinder. He set up this experiment very quickly using a horizontal microscope and an axon hanging vertically from a cannula. Within a day or two he came up with the unexpected answer that axoplasm is normally solid and that the mercury droplet does not fall at all, unless it is in axoplasm which has become liquefied as a result of damage or

A
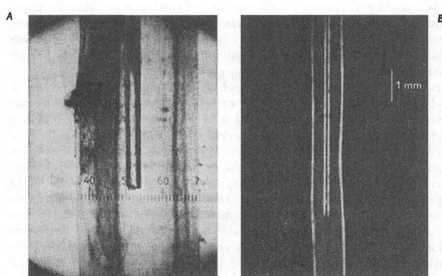
B

Fig. 5. *A*, left, photomicrograph of a recording electrode inside a giant axon, which shows as a clear space with small nerve fibres on either side; one division = 33 μm (from Hodgkin & Huxley, 1939). *B*, right, cleaned giant axon of *Loligo forbesi* with glass tube 0·1 mm in diameter inside it; dark ground illumination (from Hodgkin & Keynes, 1956).

proximity to a cut region. However, this negative experiment, which is responsible for the vertical set-up used on and off for thirty years at Plymouth, was to have an interesting sequel. Huxley said he thought it would be fairly easy to stick a capillary down the axon and record potential differences across the surface membrane. This worked at once, but we found the experiment often failed because the capillary scraped against the surface; Huxley rectified this by introducing two mirrors which allowed us to steer the capillary down the middle of the axon. Fig. 5 shows the technique. The result, illustrated in Fig. 6, was that the action potential of nearly 100 mV was about twice the resting potential of about 50 mV.

We were tremendously excited with this finding as well as with the potentialities of the technique and started other tests like the effect of potassium ions on resting potential and action potential, an experiment later done very elegantly by Curtis & Cole (1942). However, within three weeks of our first successful impalement, Hitler marched into Poland, war was declared and I had to leave the technique for eight years until it was possible to return to Plymouth in 1947.

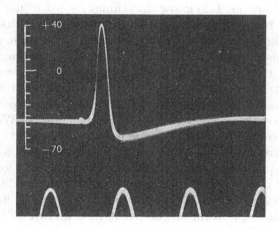

Fig. 6. Action potential and resting potential recorded between inside and outside of axon with capillary filled with sea water. Time marker 500 Hz. The vertical scale indicates the potential of the internal electrode in milli-volts, the sea water outside being taken as at zero potential (from Hodgkin & Huxley, 1939; see also Hodgkin & Huxley, 1945; Curtis & Cole, 1940).

Huxley and I wrote a cautious note to *Nature* about our results and for the first few months of the war I tried to work on a full paper. But this didn't get very far as it had to be done in the evenings after a long day at the Royal Aircraft Establishment, Farnborough, where I was working on aviation medicine with Bryan Matthews. After I had switched to radar, there were other things to study in the evenings and by June 1940 the war had gone so disastrously, and the need for centimetric radar was so pressing, that I lost all interest in neurophysiology and did not even bother to keep my copies of the *Journal of Physiology*.

By 1944 the position of the Allies had improved, radar was less demanding and I had married Peyton Rous's daughter, Marion, whom I first met in New York in 1937. There seemed to be a reasonable chance of getting back to Physiology and I was feeling happy enough to start thinking again about nerve. Cole had sent me reprints of the work which he and his group had done up to 1942 and reading these got me going again. Andrew Huxley was working on Naval gunnery and his visits to Malvern (where I worked on radar) enabled us to finish the paper about the action potential and resting potential which we had started in 1939 (Hodgkin & Huxley, 1945). Later on, both of us came to regret the discussion in that paper and I have often been asked why we did not mention the sodium hypothesis, or whether we had thought of it at that time. In the absence of documentary evidence it is dangerous to answer such questions from memory and I shall not attempt to do so here. However, I do know that things looked rather black for the sodium theory, both then and several years later. In the first place there was a report which later proved to be wrong that the action potential of *Loligo pealii* might exceed the resting potential by as much as 110 mV (Curtis & Cole, 1942); on the sodium hypothesis this required an internal sodium concentration of less than 6 mM which compared unfavourably with the value of 270 or 162 mM obtained by subtracting potassium from total base in the early analyses of Bear & Schmitt (1939) or Webb & Young (1940) respectively. There was also a preliminary report, which again proved to be wrong, to the effect that the action potential and resting potential of a squid axon were unaffected by removing all ions and circulating isosmotic dextrose (Curtis & Cole, 1942). This seemed to fit with the well-known observation that frog nerve would survive for many hours in salt-free isotonic sugar solutions, a result now known to be due to an impermeable layer in the perineurium.

I have often been asked how much working for nearly six years on radar affected the rest of my scientific life. I would like to be able to say that it made a profound difference, and I expect it did, indirectly. But the fact remains that when I returned to Cambridge in August 1945 I continued working on crustacean nerve using almost exactly the same equipment as before the war. The first big changes in technique coincided with the arrival of Richard Keynes (another ex-radar scientist), who started working on radioactive tracers after completing Physiology Part II in the summer of 1946. At the same time David Hill was developing optical methods of studying nerve and muscle and we made a push towards chemistry by bringing in Peter Lewis from the Hinshelwood school of chemistry at Oxford. All these developments, as well as scientific hospitality to a very distinguished series of visitors, were made possible by a generous grant of £3000 per annum from the Rockefeller Foundation to Professor Adrian.

But to go back to the immediate post-war period. After a few months Andrew Huxley was released from the Admiralty and we were able to continue the very happy collaboration which we had started in 1939. Some work on an indirect method of measuring potassium leakage in activity[1]

---

[1] Hodgkin & Huxley (1947). For the results of direct measurements of Na and K movements see Keynes, 1948, 1949, 1951 a, b; Keynes & Lewis, 1951; Rothenberg, 1950; Grundfest & Nachmansohn, 1950.

got us thinking quantitatively about ionic movements during the nerve impulse. Towards the end of 1946 and in the early part of 1947 we spent much time speculating about the kind of system which might give rise to an action potential.

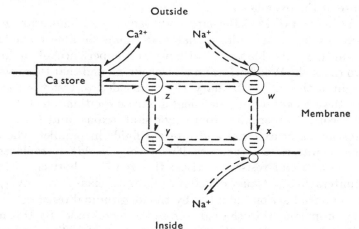

Fig. 7. Diagram illustrating theoretical carrier model considered by Hodgkin & Huxley (1948, unpublished). See also Hodgkin, Huxley & Katz (1949). With a high resting potential (inside negative), all the carrier molecules, both uncombined and combined, are pulled to the outside of the membrane and sodium movement is low; depolarization allows the carriers to move and increases the inward flow of sodium ions. The effect is increased and made asymmetric by external calcium ions which are assumed to combine with the carriers and form an immobile reservoir on the outside of the membrane.

The general idea underlying our initial hypothesis was that sodium ions were transferred across the membrane by negatively charged carrier molecules or dipoles. In the resting state these were held in one position by electrostatic forces and unable to ferry sodium ions. These carriers were subject to 'inactivation' by reacting slowly with some substances in the axoplasm. A propagated action potential calculated by Huxley in 1947 incorporated the main features that emerged two years later from the voltage-clamp experiments – i.e. a rapid rise in sodium permeability followed by a slower decay, and a slow rise of potassium permeability.

In all these theoretical action potentials the reversed potential difference at the crest of the spike depended on a selective increase in sodium permeability and a low internal concentration of sodium ions. Huxley felt all along that this was a likely mechanism but I was more doubtful, partly for the reasons given above, and partly because I hankered after a mechanism which would give a transient reversal, so accounting for repolarization, inductance and the transient nature of the spike. We tried various ingenious schemes involving dipoles that I thought might operate in this way, but Huxley's numerical calculations shot them all down, leaving a rise in permeability to sodium ions, or perhaps to an internal anion, as the most likely cause of the overshoot.

Towards the end of 1946 Bernard Katz sent me a manuscript in which he showed, among other things, that crab axons became inexcitable in salt-free solutions.[1] As this agreed with my own experience[2] of squid axons I began to think the Woods Hole result was wrong and that there was hope for the sodium theory. In January 1947 I decided to test the theory by measuring the effect of sodium-deficient dextrose solutions on (1) the action potential recorded externally from single crab axons, and (2) the longitudinal resistance of external and internal fluids in parallel. The second measurement was needed because if sodium chloride is replaced with dextrose the external resistance rises; this reduces short-circuiting and partly counteracts any 'true' effect of sodium deficiency. However, if you divide the external action potential by the longitudinal resistance you get a quantity proportional to the p.d. across the membrane. By this method I found that lowering the external concentration of sodium reduced the action potential by about the right amount – for example lowering $[Na]_0$ to $\frac{1}{5}$ reduced the action potential by 40 % – from 120 to 72 mV if one assumed equality of external and internal resistances. The reduction of 48 mV was not far from 58 log 5 = 41 mV and seemed reasonable.

These experiments were brought to an end by the first of our many energy crises, in this case precipitated by an exceptionally prolonged cold spell which lasted until the end of March 1947. It was soon found that national coal stocks were exhausted and the central heating was switched off in many buildings, including university laboratories We then had no cold room in our part of the laboratory and I remember that David Hill took the opportunity of carrying out a series of experiments at 4° C. But you can't dissect single fibres at such temperatures and I spent the time writing at home or talking with Andrew Huxley in Trinity where he could be seen cranking a Brunsviga calculating machine with mitten-covered fingers. I started experimenting again in April but by then the summer vacation was approaching and I had decided to do the sodium experiments properly at Plymouth using the squid axon and an internal electrode. I found it hard work to get going again. The Plymouth laboratory had been partly demolished in the great air raids of 1941 and was being rebuilt; squid were in short supply and I'd forgotten much of the technique. Worst of all I was short of a partner. Bernard Katz wasn't free till September and

---

[1] Katz, 1947.

[2] See small type paragraph on p. 303 of Webb & Young, 1940.

Andrew Huxley was just getting married. However, the basic experiments of determining the effect of sodium concentrations on overshoot and rate of rise are straightforward and by the time of the Oxford congress (21–25 July 1947) there was strong evidence for the sodium theory. I left the equipment at Plymouth and with the help of Bernard Katz made a thorough attack on the problem in September. We wrote up the results during the autumn but publication delays held up our paper for 15 months. Meanwhile there were interesting developments on several fronts and I shall have to depart from a strictly chronological account if I am to retain any logical order.

Since the end of the war I had corresponded fairly regularly with Kacy Cole and on 26 August 1947 I wrote to tell him about the sodium results and to discuss future joint research. We had made a tentative plan (which never came off) to join forces at Woods Hole in 1948, and were starting to discuss research possibilities. I was clearly in an optimistic mood because I wrote,

I should rather like to have a shot at perfusing the inside of the axon with potassium or sodium salts and have some ideas about the best method of doing this. I am also interested in the possibility of stimulating the axon with a diffuse electrode in such a way that the axon is excited uniformly over a length of 1 or 2 centimetres. This might give useful information about the nature of the active process uncomplicated by propagation and local circuit action. What are your plans and views?

In his reply of 7 October 1947, Cole said,

...I am sure that you will be excited to hear that we spent the whole summer with an internal electrode 15 millimetres long and about 100 microns in diameter... The two principal ideas are first the use of the central outside region with a guard region on each side, and second the use of a feedback circuit to control either the current flow in the central region or the potential difference across the membrane in that region to the desired value...

In the end the plan to spend a summer at Woods Hole fell through but I did have an exceedingly helpful visit to America in the spring of 1948 when I spent a week or two in Chicago at Cole's invitation in order to exchange information and discuss future experiments.

Andrew Huxley and I were very anxious to test our carrier theory and when we heard about Cole and Marmont's experiments we felt that voltage control, with current applied through a long metal wire, might be a good way to prove or disprove the theory. But we were worried about electrode polarization and decided to use two fine silver wires, one for current and one for voltage. Before leaving for America in March 1948 I made a short double-spiral electrode out of two 20 μm wires wound round a 100 μm glass rod. On returning to Cambridge, Mr R. H. Cook built our first feedback amplifier along roughly the same lines as those used in the final model.

Our apparatus differed in several respects from that of Cole and Marmont but it owed a great deal to the experiments which they started in 1947 and to the information which they so generously provided in the spring of 1948.

As on previous occasions we approached the experiments more circuitously than might appear from the published record. Katz and I first spent several weeks in July 1948 trying to perfuse squid axons with virtually no success, except that we learnt that calcium ions would liquefy axoplasm. Having failed here we started to try to make and insert double spiral electrodes. This didn't work either until we realized that one should first pre-drill the axon with a smooth glass capillary. Then things started to move and by using short shocks and constant currents with different external solutions we obtained indirect information about the permeability changes to sodium and potassium. Andrew Huxley arrived in mid-August and settled down to make the feed-back amplifier work. We managed to do a few voltage-clamp experiments, which were published in 1949, but realized that we needed a proper system of guard electrodes and would do better to work at low temperatures. We didn't shoot down the carrier hypothesis until the next year and for some time had no clear evidence of sodium inactivation.

During the next year Huxley and I spent a fair amount of time improving the equipment and we returned to the attack at Plymouth in June 1949. At first squid were in poor supply and we took a few weeks to get going. But by mid July 1949 Katz had joined us, there was a fine supply of living squid and in the next month we obtained virtually all the voltage-clamp records that were used in the five papers published in 1952 (Hodgkin, Huxley & Katz, 1952; Hodgkin & Huxley, 1952a, b, c, d). I think we were able to do this so quickly and without leaving too many gaps because we had spent so long thinking and making calculations about the kind of system which might produce an action potential of the kind seen in squid nerve. We also knew what we had to measure in order to reconstruct an action potential.

We spent over two years analysing and writing up the results and I have often been asked why this took so long. The answer, as usual, is multiple. In the first place we continued with some experimental research and a mild amount of teaching. The second reason which must now seem surprising is that although we had obtained much new information the overall conclusion was basically a disappointment. We had started off to test a carrier hypothesis and believed that even if that hypothesis was not correct, we should nevertheless be able to deduce a mechanism from the massive amount of electrical data that we had collected. These hopes faded as the analysis progressed. We soon realized that the carrier model could not be made to fit certain results, for example the nearly linear instantaneous current voltage relationship, and that it had to be replaced by some kind of voltage-dependent gate. As soon as we began to think about molecular mechanisms it became clear that the electrical data would by themselves yield only very general information about the class of system likely to be involved. So we settled for the more pedestrian aim of finding a simple set of mathematical equations which might plausibly represent the movement of electrically charged gating particles. But even that was not easy, as the kinetics of the conductance changes were unlike anything we had come across before, particularly the S-shaped 'on' and exponential 'off' of the conductance curves. I think we both appreciated the need to involve several particles, but it was Andrew Huxley who eventually came up with the ideas which led to the m^3h and n^4 formulation.

Finally there was the difficulty of computing the action potentials from the equations which we had developed. We had settled all the equations and constants by March 1951 and hoped to get these solved on the Cambridge University computer. However, before anything could be done we learnt that the computer would be off the air for 6 months or so while it underwent a major modification. Andrew Huxley got us out of that difficulty by solving the differential equations numerically using a hand-operated Brunsviga. The propagated action potential took about three weeks to complete and must have been an enormous labour for Andrew. But it was exciting to see it come out with the right shape and velocity and we began to feel that we had not wasted the many months that we had spent in analysing records.

In trying to give a connected account of the development which led to the voltage-clamp experiments I am conscious that I have followed only one of several interconnected strands and that I have left out a number of very important lines of research. Thus I have said nothing about the direct measurement of ionic movements with radioactive tracers and nothing about saltatory conduction or the developments which followed the introduction by Ling and Gerard of their type of micro-electrode. These omissions should not be taken to imply that these developments are of less importance. My reason for concentrating on one line is that I felt you might be more interested in informal detail than in a broader, less personal review which you can in any case obtain for yourself from the literature.

This Review Lecture was delivered at the Centenary Meeting of the Physiological Society in Cambridge in July 1976.

REFERENCES

BEAR, R. S. & SCHMITT, F. O. (1939). Electrolytes in the axoplasm of the giant nerve fibres of the squid. *J. cell. comp. Physiol.* **14**, 205–215.

BLINKS, L. R. (1930). The direct current resistance of *Nitella*. *J. gen. Physiol.* **13**, 495–508. See also Blinks (1936).

BLINKS, L. R. (1936). The effect of current flow on bioelectric potential. III. *Nitella*. *J. gen. Physiol.* **20**, 229–265.

COLE, K. S. (1941). Rectification and inductance in the squid giant axon. *J. gen. Physiol.* **25**, 29–51.

COLE, K. S. (1947). *Four Lectures on Biophysics, Rio de Janeiro*. Publication of the Institute of Biophysics, University of Brazil: Rio de Janeiro.

COLE, K. S. & CURTIS, H. J. (1938). Electric impedance of *Nitella* during activity. *J. gen. Physiol.* **22**, 37–64.

COLE, K. S. & CURTIS, H. J. (1939). Electric impedance of the squid giant axon during activity. *J. gen. Physiol.* **22**, 649–670.

COLE, K. S. & HODGKIN, A. L. (1939). Membrane and protoplasm resistance in the squid giant axon. *J. gen. Physiol.* **22**, 671–687.

CURTIS, H. J. & COLE, K. S. (1938). Transverse electric impedance of the squid giant axon. *J. gen. Physiol.* **21**, 757–765.

CURTIS, H. J. & COLE, K. S. (1940). Membrane action potentials from the squid giant axon. *J. cell. comp. Physiol.* **15**, 147–157.

CURTIS, H. J. & COLE, K. S. (1942). Membrane resting and action potentials from the squid giant axon. *J. cell. comp. Physiol.* **19**, 135–144.

GRAY, J. (1931). *A Text-book of Experimental Cytology*. Cambridge: Cambridge University Press.

GRUNDFEST, H. & NACHMANSOHN, D. (1950). Increased sodium entry into squid giant axons at high frequencies and during reversible inactivation of cholinesterase. *Fedn Proc.* **9**, 53.

HILL, A. V. (1932). *Chemical Wave Transmission in Nerve*. Cambridge: Cambridge University Press.

HODGKIN, A. L. (1936). The electrical basis of nervous conduction. Fellowship dissertation. Library of Trinity College, Cambridge.

HODGKIN, A. L. (1937a). Evidence for electrical transmission in nerve. I. *J. Physiol.* **90**, 183–210.

HODGKIN, A. L. (1937b). Evidence for electrical transmission in nerve. II. *J. Physiol.* **90**, 211–232.

HODGKIN, A. L. (1937c). A local electric response in crustacean nerve. *J. Physiol.* **91**, 5–6P.

HODGKIN, A. L. (1938). The subthreshold potentials in a crustacean nerve fibre. *Proc. R. Soc.* B **126**, 87–121.

HODGKIN, A. L. (1939). The relation between conduction velocity and the electrical resistance outside a nerve fibre. *J. Physiol.* **94**, 560–570.

HODGKIN, A. L. & HUXLEY, A. F. (1939). Action potentials recorded from inside a nerve fibre. *Nature, Lond.* **144**, 710–711.

HODGKIN, A. L. & HUXLEY, A. F. (1945). Resting and action potentials in single nerve fibres. *J. Physiol.* **104**, 176–195.

HODGKIN, A. L. & HUXLEY, A. F. (1947). Potassium leakage from an active nerve fibre. *J. Physiol.* **106**, 341–367.

HODGKIN, A. L. & HUXLEY, A. F. (1948). A theoretical model of nervous transmission. Unpublished manuscript.

HODGKIN, A. L. & HUXLEY, A. F. (1952a). Currents carried by sodium and potassium ions through the membrane of the giant axon of *Loligo*. *J. Physiol.* **116**, 449–472.

HODGKIN, A. L. & HUXLEY, A. F. (1952b). The components of membrane conduct-ance in the giant axon of *Loligo*. *J. Physiol.* **116**, 473–496.

HODGKIN, A. L. & HUXLEY, A. F. (1952c). The dual effect of membrane potential on sodium conductance in the giant axon of *Loligo*. *J. Physiol.* **116**, 497–506.

HODGKIN, A. L. & HUXLEY, A. F. (1952d). A quantitative description of membrane current and its application to conduction and excitation in nerve. *J. Physiol.* **117**, 500–544.

HODGKIN, A. L., HUXLEY, A. F. & KATZ, B. (1949). Ionic currents underlying activity in the giant axon of the squid. *Archs Sci. physiol.* **3**, 129–150.

HODGKIN, A. L., HUXLEY, A. F. & KATZ, B. (1952). Measurement of current–voltage relations in the membrane of the giant axon of *Loligo*. *J. Physiol.* **116**, 424–448.

HODGKIN, A. L. & KEYNES, R. D. (1956). Experiments on the injections of substances into the squid giant axons by means of a microsyringe. *J. Physiol.* **131**, 592–616.

HODGKIN, A. L. & RUSHTON, W. A. H. (1946). The electrical constants of a crusta-cean nerve fibre. *Proc. R. Soc.* B **133**, 444–479.

KATZ, B. (1937). Experimental evidence for a non-conducted response of nerve to subthreshold stimulation. *Proc. R. Soc.* B **124**, 244–276.

KATZ, B. (1947). The effect of electrolyte deficiency on the rate of conduction in a single nerve fibre. *J. Physiol.* **106**, 411–417.

KEYNES, R. D. (1948). The leakage of radioactive potassium from stimulated nerve. *J. Physiol.* **107**, 35P.

KEYNES, R. D. (1949). The movements of radioactive sodium during nervous activity. *J. Physiol.* **109**, 13P.

KEYNES, R. D. (1951a). The leakage of radioactive potassium from stimulated nerve. *J. Physiol.* **113**, 99–114.

KEYNES, R. D. (1951b). The ionic movements during nervous activity. *J. Physiol.* **114**, 119–150.

KEYNES, R. D. & LEWIS, P. R. (1951). The sodium and potassium content of cephalopod nerve fibres. *J. Physiol.* **114**, 151–182.

OSTERHOUT, W. J. V. (1931). Physiological studies of large plant cells. *Biol. Rev.* **6**, 369–411.

RATCLIFFE, J. A. (1975). Physics in a university laboratory before and after World War II. *Proc. R. Soc.* A **342**, 457–464, from a discussion on the effects of two world wars on the organization and development of science in the United King-dom, organized by R. V. Jones.

ROTHENBERG, M. A. (1950). Studies on permeability in relation to nerve function. II. Ionic movements across axonal membranes. *Biochim. biophys. Acta* **4**, 96–114.

RUSHTON, W. A. H. (1932). A new observation on the excitation of nerve and muscle. *J. Physiol.* **75**, 16–17P.

RUSHTON, W. A. H. (1934). A physical analysis of the relation between threshold and interpolar length in the electric excitation of medullated nerve. *J. Physiol.* **82**, 332–352.

RUSHTON, W. A. H. (1937). Initiation of the propagated disturbance. *Proc. R. Soc.* B **124**, 201–243.

SCHAEFER, H. (1936). Untersuchungen über den Muskelaktionsstrom. *Pflügers Arch. ges. Physiol.* **237**, 329–355.

WEBB, D. A. & YOUNG, J. Z. (1940). Electrolyte content and action potential of the giant nerve fibres of *Loligo*. *J. Physiol.* **98**, 299–313.

LOOKING BACK ON MUSCLE

By A. F. HUXLEY

Most of the muscle physiologists in Britain, and many in other countries, can trace their interest in muscle to the influence of A. V. Hill, whose nine-tieth birthday is due to be celebrated this year, the centenary year of the Physiological Society. In my case, part of this influence came in the usual way, by inspiration from his work and his personality, but another part of it came more indirectly. A. V.'s son David had been a little senior to me as an undergraduate at Trinity College, Cambridge, before the war; we had known each other well both at that time and during the war when, for about a year, we were working together on the application of radar to anti-aircraft (A.A.) gunnery (yet another influence from A. V. Hill, who had been a pioneer of A.A. gunnery in the 1914–1918 war, and was the main author of the *Text Book of Anti-Aircraft Gunnery*[1]). We both returned to the Physiological Laboratory of Cambridge University at the end of the war, David resuming his experimental work on muscle while I rejoined Alan Hodgkin to work on nerve. David Hill gave a course of lectures on muscle to the final-year undergraduate class, and when in 1948 he left Cambridge to take the post of Physiologist at the Marine Biological Laboratory at Plymouth, I took over this course of lectures and David gave me his lecture notes. From these I learnt, among other things, a little about the nineteenth-century work on changes in the striation pattern during contraction. I was struck by the fact that although there seemed to be perfectly good evidence for a phenomenon known as the 'reversal of striations',[2] very little had been done on it for fifty years and

[1] This comprehensive two-volume work, issued by H.M. Stationery Office in 1924–1925 for the War Office, was still a valuable reference book in the second world war. It was 'for official use only' and is not easily found in general libraries; I am grateful to Dr M. A. Hoskin, Keeper of the Archives at Churchill College, Cambridge, for allowing me to look again at a copy of it (originally A. V. Hill's own copy). The list of contributors contains at least seven who were, or subsequently became, Fellows of the Royal Society.

[2] This phenomenon was described by most of the nineteenth-century micro-scopists who observed the 'contraction waves' that pass along muscle fibres teased out from the leg muscles of insects, and that are often seen in fixed muscles from

it was not even mentioned in most textbook accounts of muscle. The phenomenon as described by Engelmann and others was that, when a fresh muscle fibre contracted under an ordinary microscope, each point which was the centre of a band with relatively low refractive index in the relaxed fibre became the centre of a narrow band with relatively high refractive index ('contraction band') when the fibre contracted to a sufficient degree.

The other factor that led me into experimental work on muscle was a long-standing interest in microscopy. When I was about ten, my parents gave me a small portable microscope, and although I studied pond life and so forth with it, I was more interested in getting a good image than in the organisms I looked at. A little later I had the use of two mid-nineteenth century microscopes that had come down from my mother's side of the family, and my brother Julian gave me the book by Drew & Wright (1927), from which I learnt some of the lore which had been known to every nineteenth-century biologist but which had been to a large extent forgotten as the microscope lost its position as the principal tool of biology. I did not take in phase contrast before the war. I remember Maurice Wilkins, another undergraduate contemporary, explaining it to me once when we were out for a walk, but without a diagram it was not intelligible to me. Shortly after my return to Cambridge in 1946, Dr A. F. W. Hughes[1] showed me his phase microscope, and I read the beautifully clear account by Zernike himself (1942). This reminded me that I had long before read somewhere of an 'interference microscope', and I devised a scheme for making one, using polarized light and with Wollaston prisms for separating the beams below the condenser and recombining them above the objective.

These two lines of thought came together in 1951, when Hodgkin and I were completing our papers on the voltage-clamp analysis of excitation in the giant nerve fibre of the squid. I felt that it was time to move into another branch of physiology: for one thing, I had never worked on anything but nerve, and for another, there was at that time (and indeed for a good many years after) no obvious way of pushing the analysis of excitation to a deeper level. A reinvestigation of the 'reversal of striations' had

many kinds of arthropods. It was clearly established during the 1870s by the papers of Flögel (1871), Merkel (1872), Engelmann (1873b, 1878), Schäfer (1873) and Fredericq (1876), though Krause (1873) claimed that the optical appearance was due to wrinkling of the sarcolemma. The 'reversal of striations' was investigated afresh in the present century by Jordan (1934), who used fixed and stained muscles from both arthropods and vertebrates. It was also seen and filmed in living tadpole tails by Speidel (1939).

[1] At the time Dr Hughes was Sir Halley Stewart Research Fellow at the Strangeways Laboratory, Cambridge. He died in 1975; there is an obituary note in the *Journal of Anatomy* 121, 399 (1976).

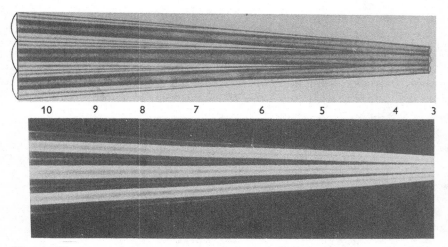

Fig. 1. Diagrams of the appearance of the striations in insect muscle at varying degrees of shortening, from Fredericq (1876). The upper part shows the appearance in ordinary light (deep focus, so that high refractive index appears dark) and the lower part shows the appearance by polarized light (no compensation, so that birefringent areas appear bright). The horizontal axis represents striation spacing, whose value in micrometres is given by the figures in the middle.

The broad dark bands in the upper diagram are the A bands, 'A' standing for 'anisotropic' since these high refractive index regions are also birefringent, i.e. optically anisotropic, as shown by the fact that they appear bright in the lower diagram. In the nineteenth century literature they are often called the 'sarcous elements' (Bowman), the 'dim bands', or the 'Q discs' (*Querscheiben*). The paler part at the centre of A is the H zone (*Hensens Streifen*; see footnote on p. 35). The black lines midway between A bands are the Z lines (*Zwischenscheiben*; see footnote on p. 34). The pale area between adjacent A bands, bisected by Z, is the I band ('isotropic' since it is hardly at all birefringent; often denoted by J in the German literature); it was also referred to as the 'clear band' in contradistinction to the 'dim' A. The grey bands within each half of the I band are the N lines (*Nebenscheiben*); these are very conspicuous in some insect muscles where they are due to regularly arranged mitochondria but faint lines of unknown composition are found in this position even when mitochondria are absent. A faint dense line at the centre of H (not shown here) is the M line (*Mittelscheibe*).

a double advantage as a subject of research: first, it was a neglected phenomenon that might give a fresh insight into the process of muscle contraction, and, second, it would give me an excuse to indulge in my hobby and
build an interference microscope.

DEVELOPING AN INTERFERENCE MICROSCOPE

The specific reasons why an interference microscope was needed are as
follows. The main feature of the striation pattern that is seen with an
ordinary microscope in a fresh muscle fibre is an alternation of bands with
higher and lower refractive index (the 'A' and 'I' bands, respectively:
see legend to Fig. 1), due to higher and lower total concentrations of
protein in these regions. There is almost no absorption of light, but the
phase differences introduced by the different refractive indices can be
made to show up by reducing the aperture of the illumination and going a
little out of focus.[1] This gives a fairly satisfactory image of muscle fibres
of small diameter and large striation spacing, such as the fibres from leg
and abdominal muscles of insects that were used in most of the nineteenth-
century investigations, but these fibres have the disadvantage that they
do not give all-or-none contractions. The contractions that the nineteenth-
century microscopists watched were the spontaneous contraction waves
which often pass along these fibres for a period after they have been
teased out, and in a contraction of this kind there is no way of telling
whether a phenomenon such as the reversal of striations is related primarily to 'activation', to tension development or to shortening. On the
other hand, Kato, and especially Ramsey & Street, had shown that it is
possible to isolate fibres from frog muscles without damage, and that these
will give all-or-none twitches or tetani and can be held at constant length
during contraction or allowed to shorten at will.[2] But these fibres are rela-

---

[1] If the specimen is defocussed by bringing the objective closer to it, an image in
'positive contrast' is obtained, i.e. high refractive index regions appear dark, but if
it is defocussed by moving the objective away from the specimen the contrast is
reversed ('negative contrast') and high refractive index regions appear bright. It
was customary to use the former position of focus (*tiefe Einstellung* in the German
literature), so that the words 'dark' (or 'dim') and 'bright' (or 'clear') were often
used as equivalent to 'more refractile' and 'less refractile', respectively. This behaviour of the image was well known to the nineteenth-century microscopists (e.g.
Krause, 1869, pp. 7–8), but an understanding of it seems to have become progres-
sively less general since 1900.

[2] Whole single fibres were isolated from frog muscles by Brown & Sichel (1930)
and by Asmussen (1932) but their preparations only gave graded responses. All-or-
none behaviour in isolated muscle fibres was, I believe, first demonstrated by Kato
(1934) and his collaborators. Buchthal and collaborators used isolated frog fibres
capable of twitches and tetani in a number of investigations before the second world

tively thick (50–150 µm) and their striations are narrow (2–3 µm repeat), so it is not usually possible to obtain a satisfactory image of the striations with an ordinary light microscope.[1] Phase contrast of the ordinary kind also cannot be relied on in a specimen of this sort because planes both above and below the plane on which the microscope is focussed give images with reversed contrast, so that the overall result in a thick striated specimen is uninterpretable. The polarizing microscope gives a fair image, but this was no good for investigating the reversal of striations, since the distribution of birefringence in the fibre did not appear to reverse in the way that the distribution of refractive index did.[2] An interference microscope which accepts a high-aperture cone of illumination, and in which the two beams are sufficiently separated for one of them to pass through the fibre while the other is clear of it, ought, however, to give a satisfactory 'optical section' in the same way that an ordinary microscope gives an optical section of a specimen in which the contrast is due to light absorption. My scheme for an interference microscope was of this kind, so I approached the firm of R. & J. Beck (now Ealing Beck Ltd) to see if they would make the necessary optical components for me, and also whether they might be interested in manufacturing the instrument commercially. The second possibility was ruled out at once: a patent on several designs of interference microscope using polarized light, including

war. A new standard of consistency and of long survival in such preparations was, however, set by R. W. Ramsey and his wife Sybil F. Street, of the Medical College of Virginia (Ramsey & Street, 1938, 1940).

[1] In a carefully dissected frog fibre at rest, places can be found where the striations are sufficiently regular and sufficiently nearly parallel to the microscope axis for a good, though not very contrasty, image of the striations to be formed. When an objective of N.A. 0·75 is used, a negative image may be formed in a narrow range of focus settings (about ± 2 µm) near the top surface of the fibre and (less often) a positive image in a similar range near the bottom of the fibre; in each case moving the plane of focus by 5 µm towards the centre of the fibre causes the image to reverse and to become more contrasty. The condenser aperture must be set to a moderate value (0·25–0·30). A striated pattern of greater contrast is seen when the microscope is focussed within the thickness of the fibre, but there is then no way of telling whether a dark band is A or I and there is no well-defined relation between the relative widths of the dark and light bands and the widths of A and I. If the condenser aperture is increased, the contrast decreases until the striations become totally invisible when the condenser aperture is about equal to that of the objective. The optical situation is probably easier in mammalian fibres, on account of their smaller diameter.

[2] Among the authors mentioned in the footnote on p. 24 the absence of a reversal in the polarized-light image was reported clearly by Flögel, Engelmann and Fredericq (see Fig. 1 in this article). Schäfer's description is not very clear. Merkel (1873) claimed that the distribution of birefringence did reverse but this observation is clearly spurious since he described the appearance as reversing when the focus was slightly changed; this does not happen in a satisfactory polarized-light image.

my scheme, had been taken out several years earlier by F. H. Smith (1947).
I am not sure whether the patent was valid, both because one of his
systems had been built in 1930 by Lebedeff and because the diagram in
the patent omitted an essential element, the analyser, but in any case
there did not appear to be a commercial future for an instrument based on
my design as Smith was already engaged on the successful development
of one of his other schemes with the firm of Charles Baker (since amalga-
mated with Cooke, Troughton & Simms to form Vickers Instruments).
However, no suitable instrument was yet available commercially, so I went
ahead with the development of a single microscope for my own use.
Messrs Beck made a pair of Wollaston prisms for me, and supplied a pair
of flat-mounted × 10 objectives (each consisting of one cemented doublet)
for use as objective and condenser. I built the mountings for the prisms,
with appropriate adjustments, and fitted them on a Swift polarizing microscope (Huxley, 1952). This instrument was easy to make, and it worked well,
within the limits set by the numerical aperture of the lenses (0·17). This,
however, was barely sufficient to resolve the striations of vertebrate muscle,
and much too small for showing details within the striation pattern.

A high-power microscope working on the same principle cannot be made
so easily, for the elementary reason that the correct places for putting the
Wollaston prisms are inside the condenser and the objective; with the
prisms just clear of the lenses, the emerging beams would be displaced
laterally instead of being coincident. I got over this difficulty by placing
above the upper Wollaston prism a sandwich of five layers of calcite
with their optical axes in different directions, which shifted one of the
beams bodily sideways by the right amount in the layers where it was the
extraordinary ray (Huxley, 1954, 1957a). Two of the other layers brought
the large path difference between the beams back to zero, so that the
instrument would work with white light. Two of the surfaces had to be
curved in order to eliminate another error due to image curvature, and
the fifth layer, with its optical axis parallel to the microscope axis, corrected a second-order error introduced by the other four. Messrs Beck again
made the optical components, which included a specially modified water
immersion objective and an oil immersion objective modified for use as a
condenser, as well as the calcite sandwich. The stand was the one manufactured by Beck for use with the reflecting objectives designed by Seeds
& Wilkins (1949), which has an additional fine focussing movement on the
stage so that the separation between objective and condenser can be left
unaltered while the specimen is moved up or down. I again made the mountings for the prisms etc. myself. I would like to record the great helpfulness of Mr R. S. Setterington (then of R. & J. Beck Ltd, now with Messrs
Degenhardt) in the development of this microscope.

OTHER INTERFERENCE MICROSCOPES

It is remarkable that a large number of interference microscopes were developed independently at about that time.[1] In addition to F. H. Smith's instruments and mine, transmission interference microscopes with two well-separated beams were made by J. Dyson (manufactured by Vickers Instruments Ltd), by J. St L. Philpot, and by G. Nomarski, and commercial instruments of the same class were also made by the West German Zeiss firm (using polarized light) and by Leitz (using duplicate optical systems). The well-known differential interference system of Nomarski (1952) is in a different category, having a very small shift between the two beams so that it shows up gradients of refractive index (or thickness) rather than the actual value of refractive index at any one point. Yet another type of interference microscope was built by Merton (1947), in which the specimen is placed between two partially silvered surfaces; this system, however, is restricted to a low numerical aperture. Tolansky (1944) developed a highly sensitive multiple-reflexion system which he and his colleagues have used extensively for study of crystal surfaces; this again is limited to low numerical apertures. Several other ingenious systems, limited to low-aperture illumination, are described by Françon (1954).

I mentioned earlier that I had a recollection of having seen a reference to an interference microscope before the second world war, but it was only after my microscope was successfully in use that I traced this to W. Ewart Williams's *Applications of Interferometry*, a very readable Methuen's Monograph that I had been given as a school prize. It mentions – though it does not actually describe – interference microscopes built by Sirks (a biologist) and by Pringsheim (a physicist) at the end of the last century. These instruments separated and recombined their beams by partial reflexion and used a single objective and condenser on a principle not unlike that of Dyson's microscope, but they were low-power instruments, using normal × 10 objectives. Williams also mentions Lebedeff's description in 1930 of a low-power interference microscope which used polarized light and which worked on a principle first devised by Jamin (1868); this was in fact the same as the principle of the 'shearing' objectives of the Smith–Baker development and of the Zeiss interference microscope. Linnik (1933) had built a two-beam reflecting system for examining surfaces, and Frederikse (1933) had built an instrument to

[1] Brief descriptions of several of the interference microscopes, both new and old, mentioned here are given, with references, in an article of mine in *Die Naturwissenschaften* (Huxley, 1957a). This article also contains micrographs illustrating the performance of my microscope.

which Merton's was almost identical. Two low-power two-beam systems were proposed by Sagnac (1911), but apparently were not actually built. All of these were more or less completely forgotten by the end of the second world war (for example, Sirks, Pringsheim and Lebedeff are not even mentioned in Françon's book (1954)), and as far as I know, all the developments of the 40s and 50s were independent fresh ideas.

It is tempting to speculate about the reason for this outburst of inventiveness to do with the light microscope. I am sure that phase contrast – the first substantial advance in microscopy since the turn of the century – was an important factor in the new interest in microscopy that certainly appeared at about that time. Given this renewed interest in microscopy, together with the emphasis on non-absorbing specimens that was also provided by phase contrast, interference microscopes were bound to appear: some ingenuity is needed in their conception and development, but they are straightforward applications of principles that were easy to understand and were well known in microscopy and interferometry. In this respect, they are very different from phase contrast itself, which depends on a fairly subtle point about the scattering of light by a weak phase specimen.

Zernike originally devised the phase contrast method as a way of testing diffraction gratings, and he says (1942) that it was independently conceived by Lyot for testing telescope mirrors. Zernike was the only person who applied it to microscopy, and the difficulty of appreciating its potential value was brought out by the lack of response when it was demonstrated in 1933 to a meeting of biologists,[1] and by the unwillingness of the Zeiss firm to undertake its development, as described in Zernike's Nobel lecture (Zernike, 1954).

MUSCLE FIBRES UNDER THE INTERFERENCE MICROSCOPE

Anyhow, the high-power version of my microscope was working satisfactorily by the end of 1952. I had been joined by Dr Rolf Niedergerke (now of the Biophysics Department, University College London), and we settled down to dissecting living single fibres from frog muscles and watching and photographing them down the microscope. A fibre surrounded by Ringer's solution does not give a satisfactory image because of the large optical path difference introduced by the fibre as a whole: not only is there a series of interference fringes parallel to the fibre axis, but the cylindrical shape of the fibre causes path differences that are different for rays travelling through the specimen in different directions. We therefore had to dissolve enough of some substance in the Ringer's solution to raise

[1] Zernike (1933). See the biographical articles on Zernike by Bleeker (1966) and by Tolansky (1967).

its refractive index approximately to the mean refractive index of the fibre. After unsuccessful attempts with egg albumin and with plasma expanders such as dextran and polyvinylpyrrolidone, we got fibres to survive well in a 20 % solution of bovine serum albumin, with appropriate salts added.

We saw very soon that when a fibre was stretched the A bands kept a more or less constant width of about $1 \cdot 5$ μm, and all the increase of length took place in the I bands. This unexpected observation immediately suggested that inextensible rodlets of this length existed in the A bands. The thing that actually suggested a sliding filament system to us was a cine film we took in March 1953 of a fibre contracting in response to a slowly increasing current. The contraction bands we were looking for did appear, but not where we expected: as the fibre shortened below its slack length the first 'contraction band' to appear was a narrow dense line at the middle of the A band, not opposite the middle of I. On more extreme shortening, however, a second set of dense lines did appear at the latter position. These bands would be nicely explained if, in addition to the rodlets needed to explain the constancy of A-band width, there was a second set of filaments in each repeat of the striation pattern, crossing the I band and overlapping with the A-band rodlets. The first set of dense lines would then be due to collision between successive sets of these I filaments, and the second set of dense lines would be due to collision between successive sets of A-band rodlets.

The difference from the nineteenth-century descriptions of contraction bands is due to the different experimental material. In arthropod muscles the first set of lines ('C_M bands') are not very conspicuous. We found, however, that they had been well described in haematoxylin-stained vertebrate muscle by Jordan (1934).

Contraction bands were not formed if the fibre was made to contract isometrically, or if it was allowed to shorten only to its normal slack length, showing that the formation of contraction bands was related to length as such, and not to 'activation' or to tension generation.

Later in 1953 we failed to repeat our observation of the formation of contraction bands in local contractions produced by steady or slowly increasing currents. We found a part of the explanation for this much later: in these circumstances contraction bands are usually formed only near the stimulated surface of the fibre, while the rest of the fibre shortens passively, the individual fibrils being thrown into waves and failing to shorten to the point where they would show contraction bands.[1] It seems that we were rather lucky to have observed their formation when we did, and there is still no explanation why that preparation in early 1953 gave

[1] Huxley & Gordon, 1962. The micrograph accompanying this note in *Nature* is so poorly reproduced that the contraction bands can hardly be seen. Frames from

contraction bands, indicating active shortening, over its whole cross-section while later preparations appeared to contract only very locally unless they gave action potentials.

At that time it had already been known for ten years[1] that the 'myosin' of earlier workers, beginning with Kühne (1864), who first described it, was a complex formed from two proteins in varying proportions: 'actin' and what is known nowadays as 'myosin'. It was natural to suppose that one of these might be the material of which the A-band rodlets were composed. In the first half of 1953 there was no evidence as to which one it was; we rather thought it was likely to be the actin because the amount of this protein contained in muscle was about right for its being the cause of the refractive index difference between A and I bands, on the basis of some rather rough estimates of this difference that we made with the interference microscope. This of course turned out to be wrong. I spent the summer of 1953 at the Marine Biological Laboratory at Woods Hole, and there I met for the first time the late Professor H. H. Weber.[2] He told me that Wilhelm Hasselbach[3] in his laboratory had recently succeeded in extracting the myosin from muscle while leaving the actin undisturbed, and had found with the electron microscope that it was the material of the A bands that had disappeared. A little later during the same visit to Woods Hole I met Hugh Huxley, who, like the late Dr Jean Hanson[4] with

---

the same cine film are satisfactorily reproduced in Plate I of the printed version of a Physiological Society Review Lecture that I gave (Huxley, 1974). A preliminary account of an electron microscope investigation into the arrangement of the filaments and their lengths, in 'wavy' fibres, was given by Brown, González-Serratos & Huxley (1970); a final account is in preparation. The apparent shortening of the bands in the passively compressed fibres is due to the obliquity of the fibrils and consequent foreshortening of A and I when the fibre as a whole is viewed transversely.

[1] Straub, 1943 (see also p. 54).

[2] At that time working in the Max-Planck Institut für medizinische Forschung, Heidelberg; he became Director of the Max-Planck Institut für Physiologie, Heidelberg, in 1954. It may be significant that twenty years earlier (Noll & Weber, 1934) he had reached the conclusion that 'myosin' (in the old sense) was localized in the A bands, on the basis of the similarities between the birefringence of those bands and the birefringence of oriented threads of 'myosin'. He died in 1974; an obituary notice by W. Hasselbach is in *Rev. Physiol. Biochem. Pharmacol.* **73**, 1–7 (1975). His daughter, Annemarie Weber, is one of the leading muscle protein chemists of the present time.   [3] Hasselbach, 1953.

[4] Jean Hanson, originally a zoologist, was one of the first members of Sir John Randall's Biophysics Research Unit of the M.R.C. at King's College London, and in 1970 became head of one of the parts into which it was divided. Apart from her very important contributions to the understanding of muscle, which began even before her collaboration with Hugh Huxley, she will be remembered for her unfailing helpfulness, especially towards junior colleagues and in the affairs of King's College. She died in 1973, at the age of 53. A biographical memoir by Randall (1975) contains many interesting historical side-lights.

whom he was working, was spending a year in the laboratory of Professor F. O. Schmitt at the Massachusetts Institute of Technology. He told me that they too had found, by examining separated myofibrils with the phase microscope, that solutions which dissolved myosin but not actin removed the dense material of the A bands.[1]

At about the same time Hugh Huxley (1953b) demonstrated, by electron microscopy of transverse sections, the double array of thick and thin filaments, which has since become familiar to everyone. The paper reporting this result contains the first proposal of the sliding-filament theory. Together with Jean Hanson, he soon afterwards found, by phase microscopy of separated myofibrils, the same constancy of A-band width, and the formation of contraction bands, that Niedergerke and I saw in whole fibres. Knowing that we were also at work on this problem he wrote to me asking whether we had an article in press to which they could refer in a note they were preparing for *Nature*, and the outcome was that we also wrote a note which appeared side by side with theirs in May 1954,[2] in addition to the communication that we were giving to the Physiological Society in March of that year. The observation of constant A-band width during shortening of myofibrils was also made at about the same time by Harman (1954).

The simple observation that it was the I bands that changed in width when a fibre was stretched or when it contracted was quite unexpected because it was the reverse of what was stated in practically every textbook at that time. As an undergraduate I had used the 1933 edition of Starling's *Principles of Human Physiology*, which quotes a paper of B. Holz (1932) as showing that the A bands shorten more than the I bands during contraction. Holz used a very thin muscle, the sternocutaneous muscle of the frog, and photographed its fibres with an ordinary light microscope.[3] A little later Fritz Buchthal and his collaborators[4] had made similar observa-

---

[1] Hanson & H. E. Huxley, 1953.

[2] H. E. Huxley & Hanson, 1954; A. F. Huxley & Niedergerke, 1954. The full account of our work appeared later in the *Journal of Physiology* (A. F. Huxley & Niedergerke, 1958).

[3] Holz does not say at what plane in the fibres he focussed, nor what condenser aperture he used; it is therefore impossible to know whether his measurements of the dark bands have any relation to the actual width of the A bands (see footnote 1, p. 26). He used an exposure of 0·1 s, which is not nearly short enough to prevent the image from being blurred by the residual longitudinal movements that occur during an 'isometric' contraction.

[4] Buchthal, Knappeis & Lindhard, 1936. These authors appreciated that a satisfactory positive image is obtained only in a narrow range of focus settings close to

tions on isolated fibres from frog muscle. and reached the same conclusion. (In a search that I made in 1954 or 1955, the only recent book I found that said anything different was the current edition of a well-known textbook which shall be nameless, and which still contained the account of the striations, due to Sharpey-Schafer,[1] which had appeared in its late nineteenth-century editions.) It was therefore natural that Niedergerke and I were surprised when we saw the A bands staying constant as we stretched our fibres. But we were almost equally surprised when, a little later in 1953, Niedergerke found a reference to a nineteenth-century report of the same observation. This was in W. Krause's book on motor endplates (1869).[2] He gives the A-band width in mammalian muscle fibres as 1·5 μm, and states that he could not detect any change in this dimension when the fibres contracted. He concluded that the A bands consist of arrays of parallel rodlets, and that during contraction these rodlets did not change in length, but fluid from the I bands moved into the A bands. He supposed that the rodlets in one sarcomere attracted those of the adjacent sarcomeres.

This stimulated me to delve further into the observations on muscle during the nineteenth century, and although this meant ploughing through a great many papers, many of them much longer than they ought to be, and most of them in German, I got a great deal of interest out of it. The first two papers which concern us, however, are in English. First, there is the famous paper of Bowman (1840),[3] which laid the foundation of our

the lower surface of the fibre. In resting fibres near slack length (striation spacing about 2·2 μm) they found the A-band width to be nearly 1·4 μm, in adequate agreement with post-war measurements. During isometric contraction, the A-band width appeared to fall to about 1·2 μm, so that A and I were nearly equal. This might easily be explained by the length of their exposure ($\frac{1}{30}$—$\frac{1}{50}$ s, still too long in relation to likely residual movements) or by vertical displacement of the fibre on stimulation, so that the position of focus was altered (see footnote, p. 26). The observations in this paper that I am unable to explain are those on resting stretched fibres, where almost all the increase of length appeared to occur in the A bands.

[1] E. A Schäfer (1850–1935), two of whose papers on muscle are referred to in this article, changed his surname to Sharpey-Schafer in 1918, out of respect to W. Sharpey who had been his teacher. He was Jodrell Professor of Physiology at University College London from 1883 to 1899, when he moved to the chair at Edinburgh. He was the only founder member of the Physiological Society who survived to its fiftieth anniversary celebrations. He is at least as well known for his work on hormones, on fat absorption, and on cerebral localization as for his investigations of muscle contraction.

[2] Krause's measurements of A-band width are on p. 12 of this book, and his statement about the absence of change in the width of the A band is on pp. 172–173. A brief description of muscle structure, with the statement that the A bands do not shorten during contraction, was published in the previous year (Krause, 1868). He upheld these views against criticisms by Engelmann in later publications (Krause, 1873, 1876; see p. 34).

[3] William (later Sir William) Bowman (1816–1892) was only 24 when this paper was published. He later became the leading ophthalmic surgeon in London. His

knowledge of muscle structure. He described how each muscle fibre – 10 to 100 μm across – is made up of fibrils about 1 μm in diameter, each of which has the alternation of denser and less dense portions, and how these fibrils are lined up accurately side by side to give the striation pattern to the whole fibre. He regarded these fibrils as the contractile structures within the fibre. I do not think, however, that he made any statement about the way the parts of the striation pattern alter during stretch or contraction. The earliest observation on this point that I have found is by W. M. Dobie (1849),[1] again examining fibrils from teased muscle fibres. He reported that the band with lower refractive index, that is to say the I band, was extremely narrow in fibrils which were slightly shortened and could sometimes be lengthened by stretching the fibril if the preparation was very fresh.

Next in chronological order comes Krause's work, which I have already mentioned. Then in quick succession there are the papers of Flögel (1871), Merkel (1872) and Engelmann (1873*b*), all using fibres from arthropods, chosen for broad striation patterns (among vertebrates there is hardly any variation in striation spacing between different muscles or between different species, while in all the arthropod classes there is wide variation so that even within a single animal it is common to find striation spacings ranging from a little over 2 μm (comparable with the vertebrate value) up to 10 or even 15 μm). All three of these authors found that moderate degrees of shortening took place by narrowing of the I bands, with relatively little change in A band width. But although they agree in this observation, their comments on it are widely different. Flögel points out that his result agrees with Krause's. Engelmann dismisses Krause's ideas on the basis that his muscle fibres (from insects) could shorten far beyond the point where successive A bands would collide with each other if they stayed at constant length; Krause (1873), however, replies by claiming that shortening beyond that point is outside the physiological range but was possible in Engelmann's experiments because his fibres were isolated and unrestrained. In more extreme degrees of shortening all three of these authors saw the formation of contraction bands, though Merkel, as I mentioned on p. 26, mistakenly claimed that the distribution of birefringence, not only the distribution of refractility, underwent a reversal.

other well-known investigations in microscopical anatomy, on the structure of the liver, the kidney (Bowman's capsule) and the mucosa of the alimentary canal, were all made in the period 1839–1842. His demonstration of the structure and function of the ciliary muscle was made simultaneously with (1847), but independently of, Ernst Brücke, whose contribution on muscle was as fundamental as Bowman's (see p. 50).

[1] In the nineteenth-century literature the structure now universally known as the Z line (*Zwischenscheibe*) was sometimes referred to as Dobie's line, and also as Krause's membrane (or *Quermembran*), and as Amici's line.

Engelmann also attacks a number of other points in Merkel's papers, mostly quantitative statements about the degree of shortening and swelling of the bands.

A few years later there are two exceptionally clear and convincing papers. One, in 1876, is by Léon Fredericq, who measured band widths in alcohol-fixed insect muscles with an eyepiece micrometer. I reproduce here as Fig. 1 his diagrams showing the changes of the band pattern as a function of striation spacing, both in ordinary light (showing the distribution of refractive index, i.e. protein concentration) and in polarized light. The A band is shown as remaining at constant width almost until it is met by the Z line, and this is also stated explicitly in the text. The 'reversal' phenomenon is clear in ordinary light but does not occur in the polarized light image; and a dense line (what would nowadays be called a C_M band) appears in the middle of A shortly before the low-refractive-index band disappears altogether. The other paper, of 1878, is by Engelmann. He first sets out very clearly the points that can be established on fresh material; he then shows that all these points are found also in fixed fibres, and gives this as grounds for confidence in further points that he establishes on the basis of fixed fibres that contain contraction waves. All his conclusions are in very close agreement with those of Fredericq, confirming the main points made in 1871–1873 by himself and by Flögel. Anyone who feels inclined to dismiss nineteenth-century observations on the striations as unreliable should read these two papers; he or she should also look at the polarized light micrograph of an insect fibre, taken by Engelmann and given to Schäfer, which is reproduced in the editions from 1907 onwards of Schäfer's *Essentials of Histology* and in Bayliss's *Principles of General Physiology* (fig. 134 in the first edition, 1915).

The only descriptions from this period that I have seen which disagree substantially with the sequence in Fig. 1 are those of Hensen (1869) and of Ranvier (1880). Both are quoted in the other papers of this time as claiming that both A and I shorten during contraction. Hensen's description is not nearly so clear as those of the other authors I have quoted, and contains a number of inconsistencies.[1] Both of them rely to a large extent on

---

[1] This is the only paper on muscle by Hensen that I have been able to trace. He describes (pp. 3–4) in mammalian muscle a thin band, to which he gives the name M (*Mittelscheibe*), in the centre of the A band. I cannot make out from his description what this was. He claims (p. 25) that Krause has confused the A and I bands with each other, and that Krause's Z is the same as his M; I think it likely that they are identical but that it was Hensen who got the A and I bands mixed up. Engelmann (1873a, pp. 36–37) clearly described the region of slightly lower refractive index in the middle of the A band in insect fibres, and generously referred to Hensen in this connexion. It is nowadays known as the H zone, H standing for Hensen. 'M' is now reserved for the very narrow, and not very conspicuous, dense line at the

observations on vertebrate muscle. Ranvier's account, which is a trans-script of lectures given in 1876, describes frog muscles fixed by injection of osmic acid, with or without tetanization (fourteenth lecture). In resting muscles he found that changes of length took place principally in the I bands (*ibid.*, p. 180), but when a stretched muscle was stimulated iso-metrically the A band became narrower. On this he based a theory in which contraction is due to a diminution in volume on the part of the A bands, while the I bands are passive elastic structures. I cannot explain his observation of the narrowing of the A bands; perhaps it was an appear-ance due to the irregularity of the striations which Ranvier himself described in that experiment. As regards insect muscle, Ranvier has a diagram (fig. 17 of Ranvier (1880)) showing I shortening much more than A in the early stages of a contraction, but he does not mention this fact in his description in the text. The paper by Schäfer (1873) is also not very clear and describes the changes in insect muscle in terms different from those used by Engelmann, but there is no substantial disagreement as far as I can see.

In a slightly later paper, Engelmann (1880) describes micrometric measurements of band widths, similar to those of Fredericq. His results are much the same, showing only a slight change of A band width until shortening is sufficient to bring the Z lines nearly into contact with its edges. The same observation is made, in qualitative terms, also by Kölliker (1888, p. 706), Retzius (1890) and Rollett (1891).

On this basis, I think it is fair to say that it was established during the 1870s that, in the muscles of arthropods, moderate degrees of shortening take place with little if any change in width of the A bands, and that in higher degrees of shortening the phenomenon described as 'reversal of the striations' occurs. There was also a strong indication (from Krause's work) that the first of these propositions was true also of vertebrate muscle. All of the authors concerned had followed Bowman in assuming that the fibrils were composed of the actual contractile material, though there was argument about the extent to which they existed as real structures in the living muscle. Further, these results received additional confirmation during the 80s and 90s from some of the most distinguished microscopists of the time. There certainly were points of disagreement, but most of these had to do either with the interpretation of the observations or with features which are variable from one fibre type to another, such as the N lines and their movements during contraction.

middle of H. Hensen did examine muscles from arthropods as well as vertebrates, but he seems not to have found any muscles with sarcomeres longer than 2·8 μm.

THE UNIFORMITY OF NATURE, AND THE RETICULAR THEORY OF MUSCLE

From the middle 1880s onwards, however, this account of striated muscle is repeatedly challenged. A factor which seems to be related is that from about the same date papers on muscle commonly contain arguments based on the principle of the Uniformity of Nature. The assumption that different forms of motility are essentially similar is indeed expressed in a tentative and legitimate way many years earlier: for example Kühne (1864) emphasizes chemical similarities between muscle and the proto-plasm of amoeboid cells, and Engelmann (1875) establishes a connexion between 'contractility' and birefringence (see also pp. 51–53), but around 1885 much more ambitious applications of the principle begin to appear. The whole emphasis of both observation and interpretation in many pieces of work on muscle during the following twenty years is clearly guided by the unquestioning assumption that the principle is universally valid.

The principle as it applies to muscular contraction is stated succinctly in the opening paragraph of a paper on the 'histology of the striped muscle fibre' by B. Melland (1885), then a young man at Owens College, Man-chester, which reads: 'Everyone who has considered the subject must admit the essential identity from a physiological point of view of all those tissues which possess in a special degree contractility. The contraction of a white blood-corpuscle or amoeba is essentially the same phenomenon as the contraction of an involuntary fibre-cell or a striped muscle-fibre.' A para-phrase of this passage, often with references to protoplasmic streaming, ciliary and flagellar movement and the stalk of *Vorticella* as well as to amoeboid motion, reappears like an incantation in numerous papers about muscle from the period 1885 to 1905. After these dogmatic assertions, it is refreshing to read the review by Biedermann (1909), in which he states his opinion that protoplasmic streaming and amoeboid movement are not comparable to muscular contraction and refers to von Ebner and Gur-witsch as adopting the same view; he discusses a number of mechanisms in addition to 'contraction' of spindle fibres as the possible basis of chromosome movements in mitosis; and he suggests that the very use of the word 'contraction' in connexion with such diverse phenomena is a reflexion of a tacit assumption that the underlying process is in all cases the same as in muscular contraction.

Several distinct uses were made of this principle of Uniformity. Melland, for example, argues that Klein[1] and others have shown that contractile

[1] E. E. Klein was a founder member of the Physiological Society; born in Slavonia (now part of Yugoslavia), he came to London in 1869 to work as Burdon Sanderson's assistant. A biographical note is to be found in Sharpey-Schafer's *History of the*

protoplasm consists of a matrix in which is embedded an intracellular network of filaments, and hence muscle should have an essentially similar
structure; he studies principally the longitudinal and transverse network
that is shown in striated muscle by staining with gold chloride, and concludes that this network is the contractile structure while the myofibrils
are artifacts and the A bands are areas of the amorphous matrix lying in
the meshes of this network. Melland's paper was followed up by a colla-
borator, Marshall (1888), who reached exactly the same conclusions.

Two papers by van Gehuchten (1886, 1888) follow the same plan as
Melland's, but on a very different scale. The first paper deals with arthropod muscle, the second with vertebrate muscle, and together they cover
nearly two hundred and fifty quarto pages. They are dedicated to van
Gehuchten's teacher at Louvain, J. B. Carnoy, whom he quotes (Carnoy,
1884) both for the proposition that protoplasm consists of a matrix
(*enchylème*) containing a fibrillar network (*réticulum*), and for the statement (attributed to Carnoy in 1880) that 'la cellule musculaire est une
cellule ordinaire dont le réticulum s'est régularisé et l'enchylème chargé
de myosine'. Van Gehuchten looks upon his own work as a development
of Carnoy's views. He again relies much on gold chloride preparations,
and regards the network as the contractile structure, with the material
of the A bands being merely an amorphous fluid containing myosin which
fills the meshes of the contractile network.

A third investigation on closely the same lines is an incursion into the
muscle field by Ramón y Cajal (1887, 1888). He acknowledges that his
approach is based on the views of Carnoy, Melland and van Gehuchten,
and he reaches exactly the same conclusion.

This reticular theory was effectively rebutted by Kölliker (1888), and
although it was supported in the same year by Kühne in his Croonian
lecture (which begins with the customary invocation of the Uniformity of
Nature), and was often referred to in the succeeding one or two decades, it
does not seem to have had any lasting influence beyond contributing to the
scepticism with which microscopic observations came to be regarded.

Neither van Gehuchten nor Cajal was a well-known figure at the time
their muscle papers were published. Van Gehuchten's muscle papers were
his first two publications. Cajal had previously published a few papers in
Spanish on topics in pathology but only one, on cell anastomoses in strati-
fied epithelia, in one of the well-known languages of science. His first
publication on the central nervous system was only in 1888, the year after

Physiological Society during its First Fifty Years. Melland refers explicitly to Klein's
Atlas of Histology, published 1879, but his views are set out in much more detail,
with references to earlier work which is mostly in the German literature, in two other
articles (Klein, 1878, 1879).

his muscle work first appeared. What seems to be his only other paper on muscle (1890) is mentioned in note 1 below. Both of these men switched their interest from muscle to the nervous system immediately after these papers; Cajal was encouraged by successes in the investigation of foetal brains with the Golgi method and persuaded van Gehuchten to follow his example.

A number of other papers about the structures between the fibrils appeared during the next two decades without implying that they were contractile and without evidence to show what function they might have (the name 'Trophospongium' used by Holmgren (1908) implies that he was thinking of a nutritive function). Apart from the mitochondria, these structures came to be completely forgotten until they were rediscovered in the 1950s with the electron microscope. This rediscovery has aroused a fresh interest in the descriptions published at the turn of the century, and also a belated admiration for them, but that is another story.[1]

In these papers by Melland, Marshall, van Gehuchten and Cajal, observations on striated muscle are given a new interpretation suggested by the principle of Uniformity. The application of the principle is taken a step further by Max Verworn (1892)[2] in his book *Die Bewegung der lebendigen*

---

[1] In most striated muscle fibres, electron micrographs show an elaborate system of tubules and vesicles in the spaces between the fibrils. These have to do with the process of turning on the contractile material which forms the fibrils themselves. Their structure and function have been worked out from 1953 onwards, but forgotten papers from the turn of the century, mostly based on the Golgi method, contained much first-rate information about them. The earliest seems to be a paper by Cajal (1890). A beautiful paper by E. Veratti (1902) was rediscovered by Professor H. S. Bennett, and an English translation of it is published as the theme of a *Supplement* to the *Journal of Biophysical and Biochemical Cytology* (**10** (1961)), together with a collection of articles on these structures. About 1959 my friend and colleague Lee Peachey wrote to the University of Pavia, where Veratti had worked, asking if a reprint of the original paper were available, and was surprised to receive a courteous reply from Veratti himself, enclosing two reprints, one of which Peachey gave to me. Veratti died in 1967; an obituary notice is in the *Rendiconti del Istituto Lombardo, Accademia di Scienze e Lettere, Parte Gen. e Atti Uff.*, **107**, 3–7 (1967). Some historical points to do with these structures are mentioned in my Croonian Lecture (Huxley, 1971), which also contains a reproduction of a figure from a paper of 1897 by G. Nyström showing that the transverse tubules are open to the external solution. The most recent reference to these old papers that I have seen, before their rediscovery in the 1950s, is in Biedermann's review (1927, fig. 15).

[2] Verworn was among the first to write a book with 'General Physiology' in its title: his *Allgemeine Physiologie* appeared first in 1894 and an English translation of its second edition appeared as *General Physiology* in 1899 (translated by F. S. Lee, and published by Macmillan). He mentions Claude Bernard's *Leçons sur les phénomènes de la vie communs aux animaux et aux végétaux* (1878) and Preyer's *Elemente der allgemeinen Physiologie* (1883) among the forerunners of this approach, and regards Johannes Müller (1801–1858) as its original proponent. Verworn also founded the *Zeitschrift für allgemeine Physiologie*, which ceased publication on his death in 1923.

Substanz, where he claims that contractility ought to be studied in its simplest manifestation, amoeboid movement, rather than in the highly specialized striated muscle fibres of arthropods and vertebrates, or even in contractile systems of intermediate degrees of specialization. Verworn deduces from observations on amoeboid movement that contraction is produced by a tendency of contractile particles to move toward the nucleus, or, if this is prevented by the structure (as in a striated muscle fibre), by an interaction between the contractile particles and 'nuclear substances' (*Kernstoffe*) which reach the A bands through the sarcoplasm and the I bands; each half-segment is said to contract because of the 'chemotropic pull of the anisotropic substance towards the isotropic layer'.

Yet another application of the principle appears in Engelmann's Croonian Lecture (1895): after the claim that 'essentially the same principle of motion' is applicable to 'all the different types of organic movement', he says, 'Only such properties and processes as all contractile structures have in common will consequently have a right to be considered essential to the process of contraction, and only such will be allowed to form the basis of our endeavours to explain muscular motion'. Engelmann's approach to the question of the essential uniformity of contractile mechanisms is altogether more satisfactory than that adopted by most of the other authors I refer to in this section, who appear to accept the principle *a priori* – probably on the basis of the triumphs of the cell theory, evolution, and comparative embryology. I shall say more about Engelmann's position on pp. 51–53.

A long theoretical paper by Julius Bernstein (1901), which develops a surface-tension theory of contraction, again contains the usual assertion of the uniformity of the nature of the contractile process.[1] He proposes (*ibid.*, p. 283) that the best structures for the study of contractility are the contractile threads in ciliates such as *Vorticella* and *Stentor*. He also draws the conclusion – which may be regarded as a special case of Engelmann's point – that, 'since the contraction process in the primitive fibrils of protozoa and in smooth muscle cells takes place, in principle, in just the same way as in the striated muscle fibre, the essential cause of the process cannot be looked for in the band structure of the latter, and all theories

[1] Bernstein published at least two more long papers (1908, 1914) supporting the surface-tension theory of contraction, but unlike his work on the origin of biological electricity, these papers have not had a lasting influence. There are interesting analogies between some of the arguments he used in these two fields: for example, he uses the rise of injury potential with temperature as a point in favour of the ionic theory of biological electricity (since an ideal concentration cell has an E.M.F. proportional to absolute temperature), and he uses the decline of twitch tension with rise of temperature as an argument in favour of the surface-tension theory of contraction (since the surface tension of an ordinary liquid declines with rise of temperature).

of contraction which are based on the layered structure of the striated fibre have little inherent probability' (p. 284).

Similar statements can be quoted from the literature of the 20s, 30s and 40s,[1] but the discoveries of the 50s, 60s and 70s have shown them to be wrong in every imaginable way: not only has the whole of the remarkable progress on contraction mechanisms since 1953 come from observations on the most highly specialized muscles, but it has been found that there are at least three types of biological motility that have little or no relationship with muscle – those associated with microtubules, with bacterial flagella, and with the contractile threads of *Vorticella* and its relatives. Actin and myosin have indeed been found to exist in numerous types of cells beside muscle, and are thought to be responsible for many forms of protoplasmic movement, but study of these 'primitive' forms of motility has contributed nothing to the understanding of muscle. The boot is on the other foot: all the current ideas about the mechanism of these forms of motility are derived from observations and experiments on vertebrate and arthropod striated muscle.

The case of the contractile thread, or 'spasmoneme', of *Vorticella* and related organisms is particularly interesting. It is this structure which causes the stem of the animal to shorten – with remarkable speed – when it is disturbed. It was found in the 1950s that the spasmoneme can be made to contract and relax by raising and lowering the concentration of calcium ions to which it is exposed,[2] and the mechanism was thoroughly elucidated by Weis-Fogh & Amos (1972). Weis-Fogh[3] rediscovered a giant colonial form, *Zoothamnium geniculatum*, related to *Vorticella*, in which the spasmoneme is large enough to be examined directly by mechanical methods. They showed that it consists of a protein with long-range elasticity of the rubbery kind, and that its natural length is decreased when it binds calcium. Relaxation is brought about simply by dissociation of this calcium when the concentration of ionized calcium in the surrounding fluid is sufficiently reduced; there is no other net chemical change, and the work

[1] An example is in Höber's *Physical Chemistry of Cells and Tissues* (1945), to which W. O. Fenn contributed a section on 'Contractility' (written in 1942). This contains the sentence (p. 447), 'The fundamental mechanism of contraction is presumably the same in all tissues, but in muscle it is less obscured by other functions...'

[2] Levine, 1956; Hoffmann-Berling, 1958.

[3] Torkel Weis-Fogh, a Dane, was one of August Krogh's last pupils. Investigations of the flight of insects led him to two highly original discoveries – the existence in many insect structures of a protein, 'resilin', with high-grade rubbery elasticity (Weis-Fogh, 1960), and a new method of generating lift, 'clap and fling', used by very small insects (Weis-Fogh, 1973). He was head of the Zoological Laboratory, Cambridge from 1966 to 1974; to the dismay of his many friends he committed suicide in November 1975.

is derived from the entropy change in transferring calcium ions from a more concentrated to a more dilute solution. Binding of calcium can be said to 'plasticize' the structure so that the material of which it is composed becomes more free to shorten under the influence of thermal motions.

Now this process is exactly what had been proposed as a theory of muscle contraction, first by Karrer (1933) and by Wöhlisch (1940) and later, in more detailed form, by Guth (1947) and by Pryor[1] (1950). Pryor in fact built a working model of 'muscle' consisting of strips of collagen (tendons from the legs of pheasants; we can imagine what happened to the rest of the experimental animals) which could be plasticized by immersion in a solution of very high ionic strength. When treated in this way they shortened, and they re-extended themselves when immersed in distilled water. As a theory of muscle, this idea lost interest when the sliding-filament process was discovered, but we are left to imagine what might have happened if Weis-Fogh's elucidation of the mechanism of the spasmoneme had come before Hugh Huxley's proposal of sliding filaments. It would have been open to people to adopt Bernstein's view that the spasmoneme is the ideal object for the study of 'contractility', and to argue that the problem of 'contractility' had been solved and that it would therefore be superfluous to look further into the details of the highly specialized contractile systems represented by the striated muscles of arthropods and vertebrates when the basic mechanism had already been discovered in a primitive organism.

Cajal's own reflexions on his paper supporting the reticular theory, written thirty years later,[2] are a lively warning against the overenthusiastic application of principles in biology:

There was current in histology at that time one of those diagrammatic conceptions which temporarily fascinate the mind and influence young workers decisively in their inquiries and opinions. I refer to the reticular theory of Heitzmann and Carnoy, which was applied very ingeniously to the constitution of the striated substance of muscles by Carnoy himself, the author of the famous *Cellular Biology*, and afterwards by the Englishman, Melland, and the Belgian, van Gehuchten. Impressed by the ability of these scientists and by the prestige of the theory, I had the weakness to regard the contractile substance, as they did, as a tiny lattice of delicate

---

[1] Mark Pryor was a few years my senior at Trinity College, Cambridge. He had a long-standing interest in high polymers; his first research was on the tanning of proteins in the cuticle of insects, and this led to war work on adhesives at the Royal Aircraft Establishment, Farnborough, and to his approach to muscle contraction. He died in 1970 as a result of a road accident.

[2] Pp. 302–304 in Cajal's autobiography *Recollections of my Life* (Ramón y Cajal, 1937). This is a fascinating book. The first half of it (originally published in 1901) is an account of his childhood as the son of a country doctor in northern Spain; his escapades outdo most of the juvenile delinquency of the present age. The passage quoted here appeared first in the original Spanish edition in 1917.

fibres (the *pre-existent filaments* seen in preparations with acids and with gold chloride) united transversely by the net postulated at the level of the line of Krause. As for the primitive fibrils, they were supposed to be the result of post mortem coagulation. Later on I changed this opinion, which was vigorously criticised by Rollet, Kölliker, and others, who declared rightly that the so-called artifacts could be observed even in the living muscles of certain insects.

I insist upon these details because I wish to warn young men against the invincible attraction of theories which simplify and unify seductively. Ruled by the theory, we who were active in histology then saw networks everywhere. What captivated us specially was that this speculation identified the complex structural substratum of the striated fibre with the simple reticulum or fibrillar framework of all protoplasm. Whatever the cell might be, amoeba or contractile corpuscle, the physiological basis or rather the active factor, was always represented by the network or elementary skeleton.

From these illusions no histologist is free, least of all the beginner. We fall into the trap all the more readily when the simple schemes stimulate and appeal to tendencies deeply rooted in our minds, the congenital inclination to economy of mental effort and the almost irresistible propensity to regard as true what satisfies our aesthetic sensibility by appearing in agreeable and harmonious architectural forms. As always, reason is silent before beauty. The case of Phryne repeats itself continually [*a footnote explains that Phryne was a Greek hetaira who, when placed on trial, won an acquittal by displaying her extraordinary beauty to the judges*]. Nevertheless, no error is useless so long as we are attended by a sincere purpose of emendation; and being convinced that enduring fame accompanies only the truth, I wished to be correct at any price. Hence, later on, I reacted vigorously against those theoretical conceptions, under which reality is lost or distorted.

NEW MICROSCOPICAL TECHNIQUES IN THE EARLY TWENTIETH CENTURY

We must now return to the specific question of how it came about that all the textbooks of the 1930s and 1940s stated that shortening took place mostly in the A bands although the constancy of A-band width during shortening of moderate degree was apparently well known and firmly established in the 1870s. First, I must mention two papers by William McDougall (1897, 1898). This is the same McDougall who was a member of A. C. Haddon's anthropological expedition to the Torres Straits in 1898–1899, and later became famous as a psychologist and philosopher at Harvard, but I think his early incursion into physiology is not widely known. Indeed a later paper by him (McDougall, 1910) is largely a complaint about the lack of recognition his first papers had received. McDougall's observations were made on 'sarcostyles', the conspicuous and easily separated myofibrils of the flight muscles of Diptera and some other orders of insects. They were repeated a few years later by Meigs (1908) with very similar results. Meigs went from America to work with Professor Biedermann, whose main concern with muscle was its excitability and mechanical properties rather than changes of internal structure, though he did write (1909) a comprehensive review of contractile mechanisms in which he supported Meigs's opinions. Now Biedermann was Professor of Physiology at Jena, and Meigs obtained help from Köhler and Ambronn at the Zeiss works. Ambronn gave him advice about birefringence, and Köhler took pictures for him with the ultraviolet microscope that he had developed.[1] These pictures, like McDougall's, showed nothing that could be called an I band, at any rate in fibrils that were prepared fresh in the saline solution. They were of uniform appearance apart from being crossed at regular intervals by thin dense lines (Z). Meigs claimed that the I bands seen in whole muscles were optical artifacts due to the high refractive index of these dense lines: as is well known, a dense structure of this kind can sometimes appear as if flanked by strips of low density on either side. His photomicrographs were first class and I see no grounds for questioning his interpretation of them. Furthermore, these separated fibrils are very much more satisfactory objects for microscopy than the large-diameter fibres which most observers had used.

Then Hürthle (1909) published a long paper based on cine micrographs of fibres from the leg muscles of insects undergoing the spontaneous contraction waves that had been studied visually by so many of his predecessors. Most of the shortening appeared to take place in the A bands, and he denied that a 'reversal of striations' occurred. Thus, the results of two impressive new techniques – ultraviolet microscopy and cine micrography – contradicted the previously established view of the changes in the striation pattern. Hürthle's paper was widely quoted, and his conclusions were supported by the papers of Holz and of Buchthal that I have already mentioned. So on purely experimental grounds it is easy to excuse the authors of the textbooks of the 1930s and 40s for disregarding the older evidence.

Causes of the discrepancies

The contradiction between Meigs's results and the observations of the 1870s is due to the use of different experimental material. It is now known that the 'fibrillar' muscles used by McDougall and by Meigs work over a very narrow range of lengths[2] and that when they are not passively ex-

---

[1] A short biography of Köhler (1866–1948) by Boegehold & Gans (1963) is to be found in *Geschichte der Mikroskopie*. Originally a zoologist, he joined the Zeiss firm in 1900 and worked for them until 1945, becoming head of their microscope section in 1938. His U.V. microscope (Köhler, 1904) used a cadmium spark (wavelength 275 nm); all its optical components were of fused quartz, and it was therefore not achromatic. It was available commercially from 1904. His name will be remembered for 'Köhler illumination', in which an image of the source is thrown by a collecting lens on to the back lens of the condenser, and for the 'Köhler compensator', now generally used for weakly birefringent specimens in the polarizing microscope.

[2] About $\pm 2\%$: Machin & Pringle (1959).

tended, the width of their I bands drops practically to zero.[1] Several of the nineteenth-century observers had examined this type of muscle and had often seen relatively broad I bands in a striation pattern similar to that of the leg muscles of insects; those preparations must have been extended beyond the normal range of lengths. In this respect, Meigs's preparations were more nearly normal, but this is no justification for his claim that I bands were always optical artifacts. In any case, extending this claim to the situation in non-fibrillar muscles was an unjustified application of the principle of Uniformity: many differences between fibrillar and non-fibrillar muscles had been repeatedly emphasized by nineteenth-century observers and were reviewed, for example, by Kölliker (1888).

The discrepancy between Hürthle's (1909) conclusions and earlier observations on the leg muscles of insects is accounted for in a somewhat similar way. The events described by Hürthle, and the appearances in his published micrographs,[2] resemble the later stages of shortening as described by Engelmann and the other nineteenth-century authors I have quoted. Even in his uncontracted fibres, the width of the I band is only about $0.7 \,\mu$m out of a sarcomere length of $6 \,\mu$m (see his tables I and V); in further shortening from this stage, the older authors had seen the formation of 'contraction bands' at the position of I, and these are non-birefringent so that the width of the 'I bands', measured with polarized light (as was done by Hürthle) would not undergo much further decrease. Hürthle does discuss, at considerable length, the possibility that his isolated fibres are already partially shortened, and mentions evidence that now seems conclusive in favour of it (*ibid.*, pp. 104–106): fibres in muscles fixed *in situ* often had striation spacings nearly double the values found in his isolated fibres at rest, and equal to the values in the extended fibres (from the same muscles and the same species) investigated by Engelmann. These long-sarcomere preparations had greatly broadened I

[1] Hodge, H. E. Huxley & Spiro, 1954.
[2] Hürthle says nothing directly about the condenser aperture used in making his micrographs, but his statements about the depth of focus (1909, p. 14) and about the illuminating arrangement in his polarizing set-up (p. 16) imply that it was very low. Most of his pictures are taken with polarized light, so that ideally a bright region should indicate the presence of birefringence, independent of the actual refractive index (related to protein concentration) at the same position. But this is achieved only if a wide condenser aperture is used, so that the striations are invisible if polarizer or analyser is removed. However, one of Hürthle's plates (his fig. 36) includes a pair of photographs of a fibre taken with polarized light at slightly different planes of focus, and the distribution of light and dark in the image reverses. This implies that he was not using a large enough condenser aperture to eliminate refraction effects. In some of his figures (e.g. fig. 12), however, the image is clearly due almost entirely to birefringence, and questions of microscopic technique appear to be less important than the questions of interpretation discussed in the text.

bands, and A bands of almost the same width as in the shorter sarcomeres of his fresh preparations. But Hürthle dismisses all sarcomeres with broad I bands as 'atypical'. This lands him in a difficulty about the actual length of the sarcomeres in his fixed extended muscles, from which he only extricates himself by supposing that the sarcomeres are labile appearances not fixed by any morphological basis (*ibid.*, p. 110) so that the total number of sarcomeres in a fibre could be less when it was fixed in the extended state than when observed fresh under zero tension. He imagines that the segmentation of a fibre into sarcomeres can vary with conditions (particularly fixation) in the same kind of way that the number of loops and nodes in an oscillating string can vary (*ibid.*, p. 113). He also found 'atypical' areas with long sarcomeres and broad I bands in passively stretched fresh fibres under the microscope (*ibid.*, table XI), as well as in those fixed preparations of extended muscles. One of the justifications he gives for dismissing these sarcomeres as 'atypical' is that the ratio of I to A in the stretched fibres was very variable, but examination of his plates (e.g. fig. 12) shows that the variability was probably due to staggering of superposed layers of fibrils – a source of error that is less easily detected in photographs than during visual observation down the miroscope, when the observer can watch the effect of focussing up and down.

Hürthle's denial of a 'reversal of striations' in contraction appears to be a purely verbal point. His descriptions of the distribution of refractive index variations along a resting fibre (*ibid.*, p. 41) and a contracted fibre (*ibid.*, p. 56) agree with Engelmann's and with what is shown here in Fig. 1, but he does not consider the difference between them to justify the word 'reversal' since in the resting fibre the densest structure is the Z line, and this is also the position of the dense part of the pattern (contraction band) in the contracted fibre (*ibid.*, p. 57). This change of emphasis may be another consequence of his avoidance of fibres extended beyond the slack length, since all observers had agreed that the I bands, with low refractive index, become more conspicuous as a fibre is stretched.

From all this it is clear to me that there is no discrepancy between Hürthle's *observations* and those of the earlier authors made on similar material. The difference lies in the choice of which of the observations to regard as relevant and important. It is rather laborious to extract this conclusion from the enormous bulk of Hürthle's paper (it is more than a hundred and sixty pages long), and I am sure that most of the later authors who quoted this paper as demolishing the older stories did not realize that they were thereby tacitly committing themselves to the idea that the sarcomeres are impermanent features whose total number in a fibre can be reduced nearly to half by applying a fixative.

THE LOSS OF INTEREST IN THE STRIATIONS IN THE TWENTIETH CENTURY

Quite apart from the question of the nature of the changes in the striation pattern during shortening, it is striking how little attention is paid at all to the striations in books and monographs dealing with muscle contraction between say 1920 and 1950. An example is Fulton's book of over 600 pages, published in 1926, where the only mentions of the striations that I have been able to find are a reference to Leeuwenhoek (*ibid.*, p. 27) and a page or so in small type (*ibid.*, p. 254). I have already mentioned several factors that must have contributed to this loss of interest in the structure of striated muscle: the argument that smooth muscles contract and therefore the striations cannot be important; the discrepancies between the main features of the accounts of the participation of different parts of the band pattern in contraction given by say Engelmann, van Gehuchten, and Hürthle; and the suggestion from Meigs's work that most of what could be seen was a set of optical artifacts. An additional factor of a different kind which probably contributed to the loss of interest in the striations was the rise of biochemistry. This, I suspect, operated in three ways. First, the striking new discoveries to do with muscle were coming from a new generation of men – Fletcher & Hopkins, A. V. Hill, Keith Lucas and Meyerhof – whose training was primarily in the physical sciences and who had not been brought up on the microscope as one of the main tools of research. I think it is generally true that a scientist does not place much reliance on conclusions drawn from techniques with which he himself is not familiar, and this lack of trust in the microscope reinforced the idea that most of the things described by the nineteenth-century microscopists were 'artifacts'.

Second, the development of a science of the colloidal state led to an appreciation of the changes of appearances in a protein gel that can be produced by the action of a fixative. W. B. Hardy (1899), in particular, carried out a thorough investigation of these changes, and concluded that 'the structure seen in cells after fixation is due to an unknown extent to the action of the fixing reagents'. The criticism was of course partially valid, especially for fixatives like mercuric chloride with which Hardy did the largest number of experiments; he did, however, use osmium tetroxide as well, and this fixative – in common use from the middle of the nineteenth century – has since been thoroughly vindicated, both by the experiments of Strangeways & Canti (1927) in which tissue culture cells viewed with dark-ground illumination were found to be almost unchanged in appearance when fixed by this substance, and by its routine use for many years in electron microscopy, which showed that it achieves a degree of preservation of detail better than anything needed with the light micro-scope. Bayliss's highly influential *Principles of General Physiology* (1915, pp. 12–17) followed Hardy's scepticism as to the appearances in fixed cells.

The third influence of the chemical approach was the emphasis on events at the molecular level, which is of course beyond the reach of the light microscope. It is no doubt true that the heart of the contraction process in muscle is a cycle of interactions between protein molecules and small molecules, and that at this level microscopy is no substitute for chemical methods, but it is one of the commonest fallacies to argue that because one approach is important, therefore another is useless. The emphasis on the chemical and physical approaches did lead to a disregard of potentially valuable clues that could have been gained from microscopy, and the theory of contraction by folding of continuous filaments, which paid no attention even to the existence of the striations and which was completely wrong, came to dominate the field for half a century.

LIGHT MICROSCOPY BETWEEN THE WARS

This lack of interest in microscope observation of muscle was widespread in the 20s and 30s but it was not complete. I have already mentioned the experiments of Holz, published in 1932. These were undertaken because G. M. Frank (1927; now head of the Institute of Biophysics of the Academy of Sciences of the USSR[1]) had published measurements of the band widths in fixed and stained frog muscle that conflicted with Hürthle's cine-micro-graphic observations. Frank's results on muscles that were immersed in fixative without stimulation are not very clear-cut but with muscles under-going isometric contractions at different degrees of stretch he found that the width of the A bands was practically constant. Frank's work was attacked on the grounds of the uncertainty of the action of fixatives (which indeed was presumably the cause of the variability of A-band width in his un-stimulated preparations), and was contradicted by Holz; it is not much quoted after Holz's paper. Another microscopist who used fixed and stained preparations was Jordan; I have already mentioned that his work helped to keep people aware of the formation of contraction bands. Buchthal (see p. 25) was among the pioneer users of muscle fibres isolated by micro-dissection, and like Holz he attempted to measure the changes in A- and I-band widths but the results in both cases were vitiated by the difficulty of getting a satisfactory image with ordinary light on a specimen which is so thick and which has such narrow striations (see note 1, p. 26). Remarkable observations by cine micrography on living muscle fibres (and also nerve fibres) in the tails of tadpoles were made by C. C. Speidel (1939)

[1] (*Added in proof*:) Professor Frank died in October 1976, aged 72.

shortly before the second world war. He found that I changed more than A in the initial stages of shortening, and that in further contraction dense lines appeared first in the middle of A and then also at the position of Z.

A few years before the second world war, Bernal[1] proposed an ingenious theory of muscle contraction which was suggested by the alternation of birefringent (A) and isotropic (I) bands along each myofibril. To this extent, it did take account of the microscopic appearance of muscle. But Bernal says nothing about the higher mean refractive index of A, which is by far the most conspicuous difference between the bands in living muscle, and which implies a concentration of protein which is higher in A than in I. And Bernal's alternating spiral arrangement of filaments within each fibril implies the reverse: where his filaments run parallel to the fibril axis, creating the birefringence of the A band, their concentration is lower than in the I bands, where they are supposed to run obliquely. This is a point which could not have escaped the microscopists of half a century before.

THE ELECTRON MICROSCOPE

One might have expected that all uncertainties about the striation pattern and its changes during stretch and contraction would have been resolved as soon as muscles were examined with the electron microscope. The real existence of the many features of the resting band pattern that had been described in the nineteenth century was indeed confirmed by the first useful electron micrographs of muscle that were published. These were micrographs of separated whole myofibrils from frog and rabbit muscle, unstained or stained with phosphotungstic acid, by Hall, Jakus & Schmitt (1946). They showed not merely the alternation of A and I bands but also Z, H, M and N. There is, however, no hint of two sets of filaments; both in this paper and in that by Rozsa, Albert Szent-Györgyi & Wyckoff (1950) it is stated that the filaments run continuously through both A and I bands. Hall, Jakus & Schmitt did find that in passive stretch there was hardly any change in length of the A bands of their fibrils, but they do not lay much stress on this point; indeed, the authors themselves do not mention it in a review published a year later,[2] and it did not attract

---

[1] Bernal, 1937. I do not think Bernal did any actual research on muscle; this article is an intriguing speculation originating no doubt in his interest in three-dimensional structure. In the 1930s there was much more interest in birefringence and its quantitative analysis, both in cellular structures and in solution, than there has been since the second world war, probably because it was then the only index of structural organization below the limit of resolution of the light microscope.

[2] Schmitt, Bear, Hall & Jakus, 1947. It was in F. O. Schmitt's laboratory at the Massachusetts Institute of Technology that H. E. Huxley and Hanson worked in 1953 (see p. 32).

attention. The same is true of other observations of constant, or nearly constant, A-band width in electron micrographs published in the late 40s and early 50s.[1] The first hint of a system of two kinds of overlapping filaments is in Hugh Huxley's (1953*b*) transverse sections.

CONTRACTILITY AND BIREFRINGENCE

The fact that a muscle fibre as a whole is birefringent was recognized by C. Boeck of Christiania in 1839,[2] but it was not until two decades later that Ernst Brücke[3] (1858) showed that this birefringence is present only (or almost only) in the bands with higher refractive index (the A bands). Brücke also measured the overall strength of the birefringence of muscle fibres stretched to varying degrees, and found that there was rather little change. This is unlike the behaviour of most materials, including fibrous biological materials such as collagen and hair, whose birefringence increases greatly with stretch, presumably by a combination of a genuine photoelastic effect and an increase in the degree of orientation of the protein chains of which they are composed. Brücke concluded that the birefringence of muscle was attributable not to continuous filaments but to submicroscopic rodlets which changed their positions relative to one another when the muscle was stretched, without themselves undergoing any change of shape. He named these particles 'disdiaclasts'.[4] There is no suggestion in his paper that each of these particles extended from one side of the A band to the other and Brücke says nothing about constancy of A-band width or indeed about changes in the width or appearance of

---

[1] Several references are given in a review of mine (Huxley, 1957*b*, p. 270).

[2] Boeck's report (1839) is in the *Proceedings* of the first meeting of a Scandinavian scientific association, and is in Norwegian. It is to be found in very few libraries in Britain, but there is an abstract in German on pp. 1–2 of *Müllers Archiv für Anatomie und Physiologie* for 1844. Boeck used tourmaline plates as polarizer and analyser, and a gypsum plate to obtain a 'sensitive tint'; most of his report is an account of the use of this equipment to determine fibre direction in a tissue. Muscle fibres are mentioned in a list of birefringent animal structures; he indicates that, in all the fibrous materials examined (presumably including muscle), the optic axis and the slow direction are parallel to the fibre axis, but he gives no further detail about muscle. A further note by Boeck in the next volume (1840) of the same series is also on birefringence but contains nothing about muscle. He promises a full paper in the *Magazin for Naturvidenskaberne* but I have not succeeded in tracing it.

[3] Brücke (1819–1892; later von Brücke) appears to have been one of the most attractive of the nineteenth-century physiologists. He was deeply interested in art; he taught anatomy to art students, and wrote a book on the physiology of colour, which aimed to do for painting what Helmholtz's more famous *Tonempfindungen* aimed to do for music. He discovered the action of the ciliary muscle simultaneously with Bowman (see. p. 33).

[4] The word 'disdiaclastic', based on Greek roots, is exactly equivalent to 'birefringent', which is based on Latin. It was in use in the seventeenth and eighteenth centuries to describe crystals such as Iceland spar.

the bands when the fibre as a whole underwent changes in length. I think he imagined that their length was much less than the A-band width. However, when Krause, ten years later, found that A-band width did not increase with stretch, and – as already mentioned – concluded that the high refractive index of A was due to rodlets of that length, he used the word 'disdiaclasts' for these rodlets.

I first met the word 'disdiaclasts' many years before I read Brücke's paper, in one of the essays by my grandfather T. H. Huxley. I remember thinking how ridiculous it was of anyone in the middle of the nineteenth century to claim to have evidence about submicroscopic particles in protoplasm, but when I did read Brücke, I realized that his argument was perfectly valid. Exactly the same argument was used nearly a century later by Ernst Fischer (1947), apparently without realizing that it had been used by Brücke.

Polarized light was regularly used by the nineteenth-century microscopists as a means of making the A bands visible, but after Brücke I have only found one, Engelmann, who lays much stress on birefringence as a property which may contribute to an understanding of the process of contraction. Two years after his first muscle papers, Engelmann (1875) published a paper entitled 'Contractilität und Doppelbrechung', the argument of which runs as follows:

1. He had already concluded in his 1873 papers that the A bands were the seat of contractility in striated muscle, on the grounds first that the volume of A increased during shortening at the expense of I and second that the A bands bulged laterally. This suggested that contractility might be generally associated with the presence of birefringent particles (1875, p. 433).

2. Engelmann had himself worked on infusoria (these were his first scientific interest and his first paper on them was published while he was still at school), and he believed that 'through the discovery of numerous transitional forms the formerly sharp boundaries between protoplasmic, ciliary and muscular movement had collapsed', with the implication that 'the molecular mechanism of movement might, in basic principle, be the same in all these cases' (ibid., p. 433).

3. It was therefore of interest to reinvestigate the question whether these other contractile structures were birefringent like the A bands of muscle, previous reports having been mostly negative (ibid., p. 433).

4. Engelmann therefore re-examines the wall of Hydra, the contractile threads in vorticellids and other ciliates, cilia and spermatozoa, contractile protoplasm, and embryonic cardiac and skeletal muscle. In all these cases he finds birefringence similar to that of striated muscle (ibid., pp. 433–459).

5. He concludes that 'contractility, wherever and in whatever form it may occur, is linked with the presence of birefringent, positive uniaxial particles whose optic axis coincides with the direction of shortening' (*ibid.*, p. 460).

6. He points out that other tissue elements with similar birefringence, such as connective tissue fibres, tend to shorten during swelling by imbibition, and suggests that a swelling process of this kind may be the mechanism of contraction (*ibid.*, p. 460).

In two of his examples, Engelmann did not detect birefringence until he had found specially suitable organisms. As regards protoplasmic movement, he failed when examining amoebae and white blood corpuscles, but did find birefringence of the usual type in the protoplasmic contractile threads of *Actinosphaerium eichhornii*. Again, he did not succeed in detecting birefringence in the contractile stalks of vorticellids until he found the giant colonial form *Zoothamnium* – the same organism that was rediscovered by Weis-Fogh and used by him and his colleagues with such success nearly a hundred years later (see p. 41). He describes the large stalk of this organism as containing a 'Muskelstreif von entsprechend colossalen Dimensionen', in which birefringence – of the usual type – was easily seen. Later he succeeded in detecting birefringence in *Vorticella* and *Carchesium*. Engelmann says nothing about any changes in birefringence when the thread of *Zoothamnium* contracts or is stretched. If he had succeeded in making observations in these conditions he would presumably have found, as Weis-Fogh & Amos did, that elongation of the relaxed thread by some 10 % is enough to double the strength of its birefringence, and that when the thread is activated its length falls by about 35 % but the birefringence falls to zero – behaviour altogether unlike the rather small changes found by Brücke (and confirmed in modern work) in striated muscle, or even the considerably larger changes found in smooth muscle (Fischer, 1944). If he had made these observations, he might have been saved from the mistake of taking the resting birefringence as evidence that the contractile mechanism was the same as in striated muscle.

Another striking observation, in which Engelmann was more fortunate, appears in his paper of 1881. This concerns the muscle fibres with 'double-oblique striation' which are found in many molluscs and some other invertebrates. He shows that the microscopic appearance of these fibres is due to the presence of 'fibrils' of high refractive index that run spirally, in both senses, along the fibre, making an angle with the fibre axis that varies between fibres and according to the degree of extension or shortening, from a few degrees up to 60° or more. These 'fibrils' are birefringent, and Engelmann shows that the axis of the (positive) birefringence coincides with the direction of the axis of the whole fibre, i.e. the direction of shortening, and not with the direction in which the 'fibril' itself is running. He concludes that each fibril is built up of birefringent particles ('Inotagmen', equivalent to Brücke's disdiaclasts) which lie with their long axes parallel to the fibre axis but are staggered relative to their neighbours so as to produce the obliquity of the fibril direction. (This has of course been confirmed with the electron microscope (Hanson & Lowy, 1961) but by that time Engelmann's observation had ceased to be widely known and the result was presented as if it was a new discovery.) Engelmann rightly claimed this observation as a remarkable confirmation of his view that there is an association between birefringence and contractility.

Engelmann held these views to the end of his life (1909), and they are conveniently set out in his Croonian Lecture (Engelmann, 1895), as well as in his article of 1906; a supplementary hypothesis which takes an increasingly important place in his discussions is that the asymmetrical swelling of the birefringent contractile structures is due to large local temperature differences produced by chemical reactions. I can see nothing of importance wrong in his observations, and his point about the association between birefringence and contractility appears valid – even on the sliding filament theory it is fair to say that 'contractility' exists only in the A band. But his two conclusions – that all kinds of biological contractility are the same, and that they are due to imbibition of fluid by an oriented structure – have turned out to be wrong.

PARALLELS IN OTHER FIELDS OF BIOLOGY

I have traced how the phenomenon that changes of length in striated muscle take place chiefly in the I bands was established in the 1870s; questioned, largely on *a priori* grounds, in the 1880s; contradicted on the basis of observations by new methods in the first decade of this century; and finally forgotten until rediscovered with the interference microscope and the phase microscope in the 1950s. We may ask ourselves, first, was this an isolated case, and second, is it a mere matter of history or is it a sequence of events from which a lesson for present-day science can be learnt?

I have made no attempt to search for other cases where correct observations were made more than, say, seventy years ago and subsequently forgotten, but I have come across enough such cases in the fields in which I have worked – nerve, muscle and microscopy – to believe that it must be something that happened on a large scale. In muscle, there is the disappearance from memory of the transverse networks seen at the turn of the century by Veratti and others, and their rediscovery with the electron microscope in the 1950s (see p. 39). There is also the old evidence from

extraction experiments[1] that 'myosin' was located in the A bands; in
spite of fresh evidence from studies on birefringence in the 1930s[2] it seems
to have been almost universally assumed in the late 40s and early 50s
that myosin was uniformly distributed over the length of the sarcomere,
and the observations by Hasselbach and by Hanson & H. E. Huxley in
1953 that proved it to be localized in the A band came as a fresh discovery.
Again, Finck (1968) has drawn attention to the extraction of what must
have been actin by Halliburton in 1887, but this was totally forgotten by
the time of its rediscovery by Straub (1943). Another example (p. 52)
is Engelmann's evidence that the filaments of double-oblique striated fibres
run parallel to the fibre axis, rediscovered with the electron microscope.

In nerve, there was Overton's (1902) discovery of the 'indispensability
of sodium (or lithium) ions for the contractility of muscle', almost forgotten by the time of Hodgkin & Katz's (1949) discovery of the role of
sodium ions in the generation of action potentials; there was Key &
Retzius's (1876) demonstration of a permeability barrier in the sheath
surrounding a nerve trunk or bundle, forgotten in the 1940s to the extent
that the ability of a whole nerve (with sheath intact) to continue working
for hours when immersed in a sodium-free solution appeared to be an objection to the sodium theory; there was the demonstration of nodes of Ranvier
in myelinated fibres of the central nervous system, made by many microscopists around the turn of the century, forgotten to the extent that the
absence of nodes in fibres of the central nervous system could be taken as
evidence against saltatory conduction in 1947 (see Huxley & Stämpfli,
1949).

In the field of microscopy, I have already mentioned (p. 28) that interference microscopes were built at the end of the last century but had been
forgotten by the time of the revival of microscopy after the second world
war.

---

[1] Probably the first experiments of this kind were made by Krause (1869, pp.
19-21), and they were repeated by many of the microscopists I have quoted.
Schipiloff & Danilewsky (1881) showed that the material extracted when the A
bands were dissolved became birefringent on drying. It seems to have been generally
accepted that Kühne's 'myosin' was the substance that gave the A bands their
high refractive index and birefringence; this is stated for example in T. H. Huxley's
book *The Crayfish* (1880, p. 186) which was written as a general introduction to
zoology. Biedermann's review on the histochemistry of muscle (1927, p. 427)
again accepts that the A band contains anisotropic rodlets of myosin.

[2] Von Muralt & Edsall (1930) and Edsall (1942) showed by flow birefringence
measurements that their myosin extracts contained filaments whose length was
roughly equal to the A-band width. Noll & Weber (1934) analysed the birefringence
of the A bands and of oriented threads of myosin, and found close quantitative
agreement. Both groups concluded that myosin exists as oriented rodlets in the
A bands.

For parallels to the excessive use of *a priori* arguments based on the Uniformity of Nature, I can mention the argument that non-myelinated nerves conduct, therefore the nodes of Ranvier cannot be of primary importance. Again, there is the history of the non-twitching muscle fibres that exist in many of the skeletal muscles of frogs: their existence as the basis of 'tonus' in these muscles was strongly suggested by the pharmacological experiments of Sommerkamp (1928) and others, and by Krüger's histological evidence in the 1930s (see Krüger, 1952) but was widely denied in the 1930s and 40s on the grounds that 'tonus' in skeletal muscles had been shown (in mammals) to be due to asynchronous twitch activity: it was not generally accepted until the experiments of Tasaki[1] in 1942–1944 and of Kuffler & Gerard (1947) in which single nerve fibres were stimulated separately, and finally the records by Kuffler & Vaughan Williams (1953) of electrical activity in fibres of the two types made with intracellular microelectrodes. An amusing twist to this piece of history is that an additional factor which discredited Krüger's histological work in the eyes of electrophysiologists was itself an unjustified application of the Uniformity principle on Krüger's part: finding a similarity in the myofibril pattern between the slow fibres of amphibians and the fibres of the soleus muscle of mammals, he concluded that the mammalian soleus must be slow in the same sense that the amphibian fibres were slow, while every electrophysiologist knew that the soleus of, for example, the cat – the muscle used in the classical work of Eccles, Katz & Kuffler (1941) on the endplate potential – gave perfectly good action potentials and twitches.

Let no one imagine that I am denying the truth, or the importance, of generalizations which may be summarized by the phrase 'the Uniformity of Nature'. What I am saying is that these generalizations, unless very carefully worded, are not of universal application, and that it is usually impossible to tell whether they apply to a particular phenomenon until that phenomenon has been thoroughly investigated and explained. 'Uniformity' is in most cases a result of common ancestry, and so is explained by the theory of evolution, but the other side of the coin is that evolution is a process which generates diversity. In the early stages of the investigation of a phenomenon, when it would help us to know whether it is

[1] Tasaki's original papers (Tasaki & Kano, 1942; Tasaki & Mizutani, 1944; Tasaki & Tsukagoshi, 1944) leave absolutely no doubt that the familiar gastrocnemius muscles of a toad are capable of two quite distinct modes of contraction, activated by different nerve fibres. It was not, however, clear until the paper of Kuffler & Vaughan Williams (1953) that these two types of contraction are carried out by different muscle fibres. Unfortunately Tasaki's papers, published in a Japanese journal during the war, did not get known in the West until many years later and even now they are not as well known as they deserve. The same is true of some of Tasaki's work on saltatory conduction, published in *Pflügers Archiv* in 1940–1942.

'essentially the same' as some other related phenomenon, we have no reliable way of telling how great are the contributions of these two contradictory aspects of the evolutionary process. We may say that the phenomenon is something fundamental and must for that reason be essentially the same wherever it is found, but this argument is merely a definition of the word 'fundamental': if it turns out that the processes are in fact different, it will be said that they were not really 'fundamental' to living matter after all. The question reduces in fact to a matter of personal judgement, which is an adequate basis for suggesting experiments but is unreliable for drawing conclusions and is harmful when it closes the investigator's mind against other possible explanations.

CONCLUSION

At the risk of becoming sententious, I will finish by trying to point to some of the lessons that may be learnt from the story that I have been telling.

I think it is clear that there are at least two places in the story where a theory was accepted with such enthusiasm that it influenced the observations that were reported. One was the episode of the 'reticular theory' in the middle 1880s (pp. 37ff), the other, the claims of continuous filaments in the early electron microscope papers, which I am sure would not have been made if it had not been for the nearly unanimous belief in the theory of contraction by shortening of actomyosin threads. I sometimes wonder whether the danger of this kind of error may be increased by the numerous symposia and conferences that are held nowadays. The same story is told at each, usually by the same people, until on the Bellman's principle ('what I tell you three times is true') it becomes impossible even to contemplate any alternative. Some degree of isolation may be as necessary for the development of new ideas as Darwin showed it to be for the evolution of new species. I often wonder whether we are all busy reinforcing each other in our belief in the present-day cross-bridge story, and whether there may be another major surprise in store for us – perhaps already lurking in the literature of the 1920s.

One wonders whether the old observation of constant A-band width might have been kept in mind if some unifying theory – in particular, a sliding-filament theory – had been proposed. I have seen no hint of a sliding-filament theory in any publication earlier than H. E. Huxley's paper (1953*b*) but it does seem to me that the idea could well have been proposed at any time from say 1873 onwards. If it had been, would it have become established, or would it have disappeared from memory in the way that Krause's theory did? In comparison with Krause's theory (that the A bands were composed of rodlets of fixed length and the rodlets in adjacent sarcomeres attracted each other), a sliding-filament theory would have had two advantages that would have been evident even in the 1870s: it would have provided a mechanism by which shortening can go beyond the point at which adjacent A bands would come into contact, so avoiding the objection that Engelmann raised against Krause's theory, and it would have avoided the need to postulate some mysterious long-range force acting across the width of the I bands. On the other hand, I do not see that it would have had a conspicuous advantage over Engelmann's theory that shortening was a result of anisodiametric swelling of the A band with abstraction of fluid from I. That theory did incorporate the observation that moderate degrees of shortening took place without much change of A-band width; it did not survive in its original form after Hürthle's paper (1909) had seemed to show all shortening taking place in A, and it is by no means evident that a sliding-filament theory would have had a better chance of survival. Although there was good evidence in those days for myosin rodlets in A, the position was very different from what it was in 1953–1954, when there was already an indication of a double set of filaments from H. E. Huxley's (1953a) low-angle X-ray diffraction work, confirmed by his transverse sections under the electron microscope (1953b), and when phase contrast and interference microscopy made it possible to see the band changes satisfactorily in vertebrate muscle, especially the formation of a contraction band at the middle of A, and the narrowing of the H zone.

How did it come about that Hürthle's results were accepted, contradicting the extensive and careful work from 1870 onwards that I have mentioned (pp. 33–36)? Kölliker, who had defended the old position in 1888 against the attack by Melland, van Gehuchten and Cajal, had died at a great age in 1905; if Engelmann had lived a little longer he would probably have criticized Hürthle's paper effectively but he died in May 1909, only a few months after its publication. Schäfer (1910) did write a brief criticism, but did not mention what seems to me to be the main cause of the discrepancy from the older descriptions, namely the fact that Hürthle dismissed as 'atypical' all fibres that were sufficiently elongated to have an I band of appreciable width (see p. 46). But I think the chief factor was that the main stream of muscle research had moved away from microscopy, and Hürthle's paper did not get scrutinized as thoroughly as it ought. Also, Hürthle's claim that only the birefringent band shortened seemed to fit with Engelmann's generalization that all formed contractile elements are birefringent.

Comparing Hürthle's paper of 1909 with those of Engelmann in the 1870s, I found myself forced to the conclusion that Engelmann had been a great deal more careful than Hürthle. Engelmann studied fibres from numerous sources but Hürthle only from one. Engelmann stressed the need to use fibres of small diameter, but Hürthle used fibres that are mostly above the limit of size that Engelmann regarded as acceptable. Then there are Hürthle's wild proposals about the transient nature of the striations, which he did not attempt to check by actual measurement or counting. I suspect this may be a case of something that happens often: the pioneer in a field is careful and thorough but his successors rush in with the idea that the techniques must be easy because they have been used before. I am reminded of a fascinating lecture that I heard given by Dr A. C. Walker of University College London, on bridge disasters. Each time a new type of bridge structure is introduced, the earliest examples are built with immense care, and they stay up, like the Brooklyn bridge (begun in 1869), the first large suspension bridge, and the Forth rail bridge (1882), for many years the biggest cantilever bridge. Then corners begin to be cut, and things are not so thoroughly checked; one of the cantilevers of the Quebec bridge collapses in 1907, and one of its centre spans is dropped into the St Lawrence in 1916; the Tacoma suspension bridge goes into an aerodynamic oscillation and breaks up a few months after its completion in 1940. And, lest anyone should think these accidents are a thing of the distant past, the same sequence has repeated itself with box-girder bridges in the last two decades.

REFERENCES

ASMUSSEN, E. (1932). Ueber die Reaktion isolierter Muskelfasern auf direkte Reize. *Pflügers Arch. ges. Physiol.* 230, 263–272.

BAYLISS, W. M. (1915). *Principles of General Physiology.* London: Longmans, Green.

BERNAL, J. D. (1937). A speculation on muscle. In *Perspectives in Biochemistry*, ed. NEEDHAM, J. & GREEN, D. E., pp. 45–65. London: C.U.P.

BERNSTEIN, J. (1901). Die Energie des Muskels als Oberflächenenergie. *Pflügers Arch. ges. Physiol.* 85, 271–312.

BERNSTEIN, J. (1908). Zur Thermodynamik der Muskelkontraktion. 1. Ueber die Temperaturkoeffizienten der Muskelenergie. *Pflügers Arch. ges. Physiol.* 122, 129–195.

BERNSTEIN, J. (1914). Zur physikalisch-chemischen Analyse der Zuckungskurve des Muskels. *Pflügers Arch. ges. Physiol.* 156, 299–313.

BIEDERMANN, W. (1909). Vergleichende Physiologie der irritablen Substanzen. *Ergebn. Physiol.* 8, 26–211.

BIEDERMANN, W. (1927). Histochemie der quergestreiften Muskelfasern. *Ergebn. Biol.* 2, 416–504.

BLEEKER, C. E. (1966). F. Zernike, 1888–1966. In *Geschichte der Mikroskopie*, vol. III, ed. FREUND, H. & BERG, A., pp. 509–513. Frankfurt am Main: Umschau Verlag.

BOECK, C. (1839). Bemaerkninger, oplyste ved Afbildninger, angaaende Anvendelsen af polariseret Lys ved mikroskopiske Undersögelser af organiske Legemer. *Förhandlingar Skand. Naturforskare möte, Götheborg 1839*, pp. 107–112.

BOEGEHOLD, H. & GANS, H. (1963). A. Köhler, 1866–1948. In *Geschichte der Mikroskopie*, vol. I, ed. FREUND, H. & BERG, A., pp. 235–243. Frankfurt am Main: Umschau Verlag.

BOWMAN, W. (1840). On the minute structure and movements of voluntary muscle. *Phil. Trans.* pp. 457–501.

BROWN, D. E. S. & SICHEL, F. J. M. (1930). The myogram of the isolated skeletal muscle cell. *Science*, 72, 17–18.

BROWN, L. M., GONZÁLEZ-SERRATOS, H. & HUXLEY, A. F. (1970). Electron microscopy of frog muscle fibres in extreme passive shortening. *J. Physiol.* 208, 86–88P.

BRÜCKE, E. (1858). Untersuchungen über den Bau der Muskelfasern mit Hülfe des polarisirten Lichtes. *Denkschr. Akad. Wiss. Wien, math.-naturwiss. Kl.* 15, 69–84.

BUCHTHAL, F., KNAPPEIS, G. G. & LINDHARD, J. (1936). Die Struktur der quergestreiften, lebenden Muskelfaser des Frosches in Ruhe und während der Kontraktion. *Skand. Arch. Physiol.* 73, 163–198.

CARNOY, J. B. (1884). *Biologie cellulaire*, Fasc. 1, pp. 192–193 & Fig. 38. Lierre: van In.

DOBIE, W. M. (1849). Observations on the minute structure and mode of contraction of voluntary muscle fibre. *Ann. Mag. nat. Hist.* ser 2, 3, 109–119.

DREW, A. H. & WRIGHT, L. (1927). *The Microscope: A Practical Handbook.* London: Religious Tract Society.

ECCLES, J. C., KATZ, B. & KUFFLER, S. W. (1941). Nature of the 'endplate potential' in curarized muscle. *J. Neurophysiol.* 4, 362–387.

EDSALL, J. T. (1942). Streaming birefringence and its relation to particle size and shape. *Adv. colloid Sci.* 1, 269–316.

ENGELMANN, T. W. (1873a). Mikroskopische Untersuchungen über die quergestreifte Muskelsubstanz. Erster Artikel. *Pflügers Arch. ges. Physiol.* 7, 33–71.

ENGLEMANN, T. W. (1873b). Mikroskopische Untersuchungen über die quergestreifte Muskelsubstanz. Zweiter Artikel. Die thätige Muskelsubstanz. *Pflügers Arch. ges. Physiol.* 7, 155–188.

ENGELMANN, T. W. (1875). Contractilität und Doppelbrechung. *Pflügers Arch. ges. Physiol.* 11, 432–464.

ENGELMANN, T. W. (1878). Neue Untersuchungen über die mikroskopischen Vorgänge bei der Muskelcontraktion. *Pflügers Arch. ges. Physiol.* 18, 1–25.

ENGELMANN, T. W. (1880). Mikrometrische Untersuchungen an contrahirten Muskelfasern. *Pflügers Arch. ges. Physiol.* 23, 571–590.

ENGELMANN, T. W. (1881). Ueber den faserigen Bau der contractilen Substanzen, mit besonderer Berücksichtigung der glatten und doppelt schräggestreiften Muskelfasern. *Pflügers Arch. ges. Physiol.* 25, 538–565.

ENGELMANN, T. W. (1895). On the nature of muscular contraction. *Proc. R. Soc.* 57, 411–433.

ENGELMANN, T. W. (1906). Zur Theorie der Contractilität. *Sitzber. Kön. Preuss. Akad. Wiss.* 39, 694–724.

FINCK, H. (1968). On the discovery of actin. *Science*, 160, 332.

FISCHER, E. (1944). The birefringence of striated and smooth mammalian muscles. *J. cell. comp. Physiol.* 23, 113–130.

FISCHER, E. (1947). Birefringence and ultrastructure of muscle. *Ann. N.Y. Acad. Sci.* 47, 783–797 (p. 787).

FLÖGEL, J. H. L. (1871). Ueber die quergestreiften Muskeln der Milben. *Arch. mikr. Anat.* 8, 69–80.

FRANÇON, M. (1954). *Le microscope à contraste de phase et le microscope interférentiel.* Paris: CNRS.

FRANK, G. (1927). Das histologische Bild der Muskelkontraktion. *Pflügers Arch. ges. Physiol.* **218**, 37–53.

FREDERICQ, L. (1876). Note sur la contraction des muscles striés de l'Hydrophile. *Bull. Acad. roy. Belg.* ser. 2, **41**, 583–594.

FREDERIKSE, A. M. (1933). Mikroskopische Beobachtung lebender Zellen. *Acta brevia Neerl.* **3**, 121–122.

FULTON, J. F. (1926). *Muscular Contraction and the Reflex Control of Movement.* London: Baillière, Tindall & Cox.

VAN GEHUCHTEN, A. (1886). Étude sur la structure intime de la cellule musculaire striée. *La Cellule*, **2**, 289–453 + I–V.

VAN GEHUCHTEN, A. (1888). Étude sur la structure intime de la cellule musculaire striée chez les vertébrés. *La Cellule*, **4**, 245–316 + I–II.

GUTH, E. (1947). Muscular contraction and rubberlike elasticity. *Ann. N.Y. Acad. Sci.* **47**, 715–766.

HALL, C. E., JAKUS, M. A. & SCHMITT, F. O. (1946). An investigation of cross striations and myosin filaments in muscle. *Biol. Bull.* **90**, 32–50.

HALLIBURTON, W. D. (1887). On muscle-plasma. *J. Physiol.* **8**, 133–202.

HANSON, J. & HUXLEY, H. E. (1953). Structural basis of the cross-striations in muscle. *Nature, Lond.* **172**, 530–532.

HANSON, J. & LOWY, J. (1961). The structure of the muscle fibres in the translucent part of the adductor of the oyster *Crassostrea angulata. Proc. R. Soc.* B **154**, 173–196.

HARDY, W. B. (1899). On the structure of cell protoplasm. *J. Physiol.* **24**, 158–210.

HARMAN, J. W. (1954). Contractions of skeletal muscle myofibrils by phase microscopy. *Fedn Proc.* **13**, 430.

HASSELBACH, W. (1953). Elektronenmikroskopische Untersuchungen an Muskelfibrillen bei totaler und partieller Extraktion des L-Myosins. *Z. Naturforsch.* **8b**, 449–454.

HENSEN, V. (1869). Ueber ein neues Strukturverhältniss der quergestreiften Muskelfaser. *Arb. Kieler physiol. Inst. 1868*, pp. 1–26.

HÖBER, R. (1945). *Physical Chemistry of Cells and Tissues.* London: Churchill.

HODGE, A. J., HUXLEY, H. E. & SPIRO, D. (1954). Electron microscope studies on ultrathin sections of muscle. *J. exp. Med.* **99**, 201–206.

HODGKIN, A. L. & KATZ, B. (1949). The effect of sodium ions on the electrical activity of the giant axon of the squid. *J. Physiol.* **108**, 37–77.

HOFFMANN-BERLING, H. (1958). Der Mechanismus eines neuen, von der Muskelkontraktion verschiedenen Kontraktionszyklus. *Biochim. biophys. Acta* **27**, 247–255.

HOLMGREN, E. (1908). Ueber die Trophospongien der quergestreiften Muskelfasern, nebst Bemerkungen über den allgemeinen Bau dieser Fasern. *Arch. mikr. Anat.* **71**, 165–247.

HOLZ, B. (1932). Die Struktur der überlebenden quergestreiften Muskelfaser des Frosches während der Kontraktion. *Pflügers Arch. ges. Physiol.* **230**, 246–254.

HÜRTHLE, K. (1909). Ueber die Struktur der quergestreiften Muskelfasern von Hydrophilus im ruhenden und tätigen Zustand. *Pflügers Arch. ges. Physiol.* **126**, 1–164.

HUXLEY, A. F. (1952). Applications of an interference microscope. *J. Physiol.* **117**, 52–53P.

HUXLEY, A. F. (1954). A high-power interference microscope. *J. Physiol.* **125**, 11–13P.

HUXLEY, A. F. (1957a). Das Interferenz-Mikroskop und seine Anwendung in der biologischen Forschung. *Naturwissenschaften, 1957*, Heft 7, pp. 189–196.

HUXLEY, A. F. (1957b). Muscle structure and theories of contraction. *Prog. Biophys. biophys. Chem.* **7**, 255–318.

HUXLEY, A. F. (1971). The activation of muscle and its mechanical response (Croonian Lecture). *Proc. R. Soc.* B **178**, 1-27.

HUXLEY, A. F. (1974). Muscular contraction (Review Lecture). *J. Physiol.* **243**, 1-43.

HUXLEY, A. F. & GORDON, A. M. (1962). Striation patterns in active and passive shortening of muscle. *Nature, Lond.* **193**, 280–281.

HUXLEY, A. F. & NIEDERGERKE, R. (1954). Interference microscopy of living muscle fibres. *Nature, Lond.* **173**, 971-973.

HUXLEY, A. F. & NIEDERGERKE, R. (1958). Measurement of the striations of isolated muscle fibres with the interference microscope. *J. Physiol.* **144**, 403–425.

HUXLEY, A. F. & STÄMPFLI, R. (1949). Evidence for saltatory conduction in peripheral myelinated nerve fibres. *J. Physiol.* **108**, 315–339.

HUXLEY, H. E. (1953a). X-ray analysis and the problem of muscle. *Proc. R. Soc.* B **141**, 59–62.

HUXLEY, H. E. (1953b). Electron microscope studies of the organisation of the filaments in striated muscle. *Biochim. biophys. Acta* **12**, 387–394.

HUXLEY, H. E. & HANSON, J. (1954). Changes in the cross-striations of muscle during contraction and stretch and their structural interpretation. *Nature, Lond.* **173**, 973–976.

HUXLEY, T. H. (1880). *The Crayfish. An Introduction to the Study of Zoology*, p. 186. London: Kegan Paul.

JAMIN, M. J. (1868). Sur un réfracteur différentiel pour la lumière polarisée. *C. r. hebd. Séanc. Acad. Sci., Paris* **67**, 814–816.

JORDAN, H. E. (1934). Structural changes during contraction in the striped muscle of the frog. *Am. J. Anat.* **55**, 117–133.

KARRER, E. (1933). Kinetic theory of muscular contraction. *Protoplasma* **18**, 475–489.

KATO, G. (1934). The microphysiology of nerve. Tokyo: Maruzen.

KEY, A. & RETZIUS, G. (1876). *Studien in der Anatomie des Nervensystems und des Bindegewebes*, vol. **2**, pp. 102–112. Stockholm: Samson & Wallin.

KLEIN, E. (1878, 1879). Observations on the structure of cells and nuclei. I, *Quart. J. micr. Sci. N.S.* **18**, 315–339; II, *Quart. J. micr. Sci. N.S.* **19**, 125–175.

KÖHLER, A. (1904). Mikrophotographische Untersuchungen mit ultraviolettem Licht. *Z. wiss. Mikrosk.* **21**, 129–165 and 273–304.

KÖLLIKER, A. v. (1888). Zur Kenntnis der quergestreiften Muskelfasern. *Z. wiss. Zool.* **47**, 689–710.

KRAUSE, W. (1868). Ueber den Bau der quergestreiften Muskelfaser. *Z. rationeller Med.* **33**, 265–270.

KRAUSE, W. (1869). *Die motorischen Endplatten der quergestreiften Muskelfasern.* Hannover: Hahn.

KRAUSE, W. (1873). Die Contraction der Muskelfaser. *Pflügers Arch. ges. Physiol.* **7**, 508–514.

KRAUSE, W. (1876). *Handbuch der menschlichen Anatomie*, 3rd edn., vol. **1**, p. 92. Hannover: Hahn.

KRÜGER, P. (1952). *Tetanus und Tonus der quergestreiften Skelettmuskeln der Wirbeltiere und des Menschen.* Leipzig: Geest & Portig.

KUFFLER, S. W. & GERARD, R. W. (1947). The small-nerve motor system to skeletal muscle. *J. Neurophysiol.* **10**, 383–394.

KUFFLER, S. W. & VAUGHAN WILLIAMS, E. M. (1953). Small-nerve junctional potentials. The distribution of small motor nerves to frog skeletal muscle, and the membrane characteristics of the fibres they innervate. *J. Physiol.* **121**, 289–317.

KÜHNE, W. (1864). *Untersuchungen über das Protoplasma und die Contractilität.* Leipzig: Engelmann.

KÜHNE, W. (1888). On the origin and causation of vital movement. *Proc. R. Soc.* **44**, 427–448.

LEVINE, L. (1956). Contractility of glycerinated Vorticellae. *Biol. Bull. Woods Hole,* **111**, 319.

LINNIK, W. (1933). Ein Apparat für mikroskopisch-interferometrische Untersuchung reflektierender Objekte (Mikrointerferometer). *C. r. Acad. Sci. URSS* **1**, 21–23.

McDOUGALL, W. (1897). On the structure of cross-striated muscle, and a suggestion as to the nature of its contraction. *J. Anat. Physiol.* **31**, 410–441 and 539–585.

McDOUGALL, W. (1898). A theory of muscular contraction. *J. Anat. Physiol.* **32**, 187–210.

McDOUGALL, W. (1910). A note in reply to criticisms of my theory of muscular contraction. *Quart. J. exp. Physiol.* **3**, 53–62.

MACHIN, K. E. & PRINGLE, J. W. S. (1959). The physiology of insect fibrillar muscle. II. Mechanical properties of a beetle flight muscle. *Proc. R. Soc.* B **151**, 204–225.

MARSHALL, C. F. (1888). Observations on the structure and distribution of striped and unstriped muscle in the animal kingdom, and a theory of muscular contraction. *Quart. J. micr. Sci.* **28**, 75–107.

MEIGS, E. B. (1908). The structure of the element of cross-striated muscle, and the changes of form which it undergoes during contraction. *Z. allg. Physiol.* **8**, 81–120.

MELLAND, B. (1885). A simplified view of the histology of the striped muscle fibre. *Quart. J. micr. Sci.* **25**, 371–390.

MERKEL, F. (1872). Der quergestreifte Muskel. I. Das primitive Muskelelement der Arthropoden. *Arch. mikr. Anat.* **8**, 244–268.

MERKEL, F. (1873). Der quergestreifte Muskel. II. Der Contractionsvorgang im polarisirten Licht. *Arch. mikr. Anat.* **9**, 293–307.

MERTON, T. (1947). On a method of increasing contrast in microscopy. *Proc. R. Soc.* A **189**, 309–313.

VON MURALT, A. & EDSALL, J. T. (1930). Studies in the physical chemistry of muscle globulin. III. The anisotropy of myosin and the angle of isocline. *J. biol. Chem.* **89**, 315–350.

NOLL, D. & WEBER, H. H. (1934). Polarisationsoptik und molekularer Feinbau der Q-Abschnitte des Froschmuskels. *Pflügers Arch. ges. Physiol.* **235**, 234–246.

NOMARSKI, G. (1952). Brevet français 1 059 123.

NYSTRÖM, G. (1897). Ueber die Lymphbahnen des Herzens. *Arch. Anat. Physiol.* (*Anat. Abt.*) pp. 361–378.

OVERTON, E. (1902). Beiträge zur allgemeinen Muskel- und Nervenphysiologie. II. Mittheilung. Ueber die Unentbehrlichkeit von Natrium- (oder Lithium-) Ionen für den Contractionsact des Muskels. *Pflügers Arch. ges. Physiol.* **92**, 346–386.

PRYOR, M. G. M. (1950). Mechanical properties of fibres and muscles. *Prog. Biophys. biophys. Chem.* **1**, 216–268.

RAMÓN Y CAJAL, S. (1887). *Bol. Inst. med. Valenciano,* 1887. Cited by Ramón y Cajal (1888).

RAMÓN Y CAJAL, S. (1888). Observations sur la texture des fibres musculaires des pattes et des ailes des insectes. *Int. J. Anat. Physiol.* **5**, 205–232 and 253–276.

RAMÓN Y CAJAL, S. (1890). Coloration par la méthode de Golgi des terminaisons des trachées et des nerfs dans les muscles des ailes des insectes. *Z. wiss. Mikrosk.* **7**, 332–342.

LOOKING BACK ON MUSCLE 63

RAMÓN Y CAJAL, S. (1937). *Recollections of My Life*, transl. E. H. Craigie. Philadelphia: American Philosophical Society.

RAMSEY, R. W. & STREET, S. F. (1938). The alpha excitability of the local and propagated mechanical response in isolated single muscle fibers. *J. cell. comp. Physiol.* **12**, 361–378.

RAMSEY, R. W. & STREET, S. F. (1940). The isometric length-tension diagram of isolated skeletal muscle fibers of the frog. *J. cell. comp. Physiol.* **15**, 11–34.

RANDALL, SIR JOHN (1975). Emmeline Jean Hanson. *Biogr. Mem. Fellows R. Soc. Lond.* **21**, 313–344.

RANVIER, L. (1880). *Leçons d'anatomie générale sur le système musculaire.* Paris: Delahaye.

RETZIUS, G. (1890). *Muskelfibrille und Sarcoplasma*, p. 72. *Biol. Untersuchungen, N.F.* **1**, 51–88.

ROLLETT, A. (1891). Untersuchungen über Contraction und Doppelbrechung der quergestreiften Muskelfasern. *Denkschr. Akad. Wiss. Wien, math.-naturwiss. Kl.* **58**, 41–98 (see pp. 61, 69).

ROZSA, G., SZENT-GYÖRGYI, A. & WYCKOFF, R. W. G. (1950). The fine structure of myofibrils. *Expl Cell Res.* **1**, 194–205.

SAGNAC, G. (1911). Strioscope et striographe interférentiels. Forme interférentielle de la méthode des stries. *Le Radium* **8**, 241–253.

SCHÄFER, E. A. (1873). On the minute structure of the leg-muscles of the water beetle. *Phil. Trans.* **163**, 429–443.

SCHÄFER, E. A. (1910). On McDougall's theory of muscular contraction, with some remarks on Hürthle's observations on muscle structure and the changes which it undergoes in contraction. *Quart. J. exp. Physiol.* **3**, 63–74.

SCHIPILOFF, C. & DANILEWSKY, A. (1881). Ueber die Natur der anisotropen Substanzen des quergestreiften Muskels und ihre Vertheilung im Muskelbündel. *Hoppe-Seyl. Z.* **5**, 349–365.

SCHMITT, F. O., BEAR, R. S., HALL, C. E. & JAKUS, M. A. (1947). Electron microscope and X-ray diffraction studies of muscle structure. *Ann. N.Y. Acad. Sci.* **47**, 799–812.

SEEDS, W. E. & WILKINS, M. H. F. (1949). A simple reflecting microscope. *Nature, Lond.* **164**, 228–229.

SMITH, F. H. (1947). British provisional patent specification no: 21996.

SOMMERKAMP, H. (1928). Das Substrat der Dauerverkürzung am Froschmuskel. *Arch. exp. Path. Pharmak.* **128**, 99–115.

SPEIDEL, C. C. (1939). Studies of living muscles. II. Histological changes in single fibers of striated muscle during contraction and clotting. *Am. J. Anat.* **65**, 471–529.

STRANGEWAYS, T. S. P. & CANTI, R. G. (1927). The living cell *in vitro* as shown by dark-ground illumination and the changes induced in such cells by fixing reagents. *Quart. J. micr. Sci. N.S.* **71**, 1–14.

STRAUB, F. B. (1943). Actin. *Stud. Inst. med. Chem. Univ. Szeged (1942)* **2**, pp. 3–15.

TASAKI, I. & KANO, H. (1942). Isolation of slow motor fiber. *Jap. J. med. Sci.* (III *Biophys.*) **9**, 17*.

TASAKI, I. & MIZUTANI, K. (1944). Comparative studies on the activities of the muscle evoked by two kinds of motor nerve fibres. Part I. Myographic studies. *Jap. J. med. Sci.* (III *Biophys.*) **10**, 237–244.

TASAKI, I. & TSUKAGOSHI, M. (1944). Comparative studies on the activities of the muscle evoked by two kinds of motor nerve fibres. Part II. The electromyogram. *Jap. J. med. Sci.* (III *Biophys.*) **10**, 245–251.

TOLANSKY, S. (1944). New contributions to interferometry. Part II. New interference phenomena with Newton's rings. *Phil. Mag.* ser. 7, **35**, 120–136.

TOLANSKY, S. (1967). Frits Zernike 1888–1966. *Biogr. Mem. Fellows R. Soc. Lond.* **13**, 393–402.

VERATTI, E. (1902). Ricerche sulla fine struttura della fibra muscolare striata. *Memorie Ist. lomb. Sci. Lett.* (Cl. Sci. math. nat.) **19**, 87–133.

VERWORN, M. (1892). *Die Bewegung der lebendigen Substanz. Eine vergleichend-physiologische Untersuchung der Contractionserscheinungen.* Jena: Fischer.

WEIS-FOGH, T. (1960). A rubber-like protein in insect cuticle. *J. exp. Biol.* **37**, 889–907.

WEIS-FOGH, T. (1973). Quick estimates of flight fitness in hovering animals, including novel mechanisms for lift production. *J. exp. Biol.* **59**, 169–230.

WEIS-FOGH, T. & AMOS, W. B. (1972). Evidence for a new mechanism of cell motility. *Nature, Lond.* **236**, 301–304.

WILLIAMS, W. E. (1930). *Applications of Interferometry*, p. 44. London: Methuen.

WÖHLISCH, E. (1940). Muskelphysiologie vom Standpunkt der kinetischen Theorie der Hochelastizität und der Entspannungshypothese des Kontraktionsmechanismus. *Naturwissenschaften*, **28**, 305–312 and 326–335.

ZERNIKE, F. (1933). Een nieuwe methode van microscopische waarneming. *Hand. 24 nat. geneesk. Congres*, pp. 100–102.

ZERNIKE, F. (1942). Phase contrast, a new method for the microscopic observation of transparent objects. *Physica* **9**, 686–693 and 974–986.

ZERNIKE, F. (1954). How I discovered phase contrast. In *Les Prix Nobel en 1953*, pp. 107–114. Stockholm: Norstedt. Reprinted (1964) in *Nobel Lectures, Physics, 1942–1962*, Amsterdam: Elsevier.

THE EARLY HISTORY OF SYNAPTIC AND NEUROMUSCULAR TRANSMISSION BY ACETYLCHOLINE: REMINISCENCES OF AN EYE WITNESS

By W. FELDBERG

DID IT REALLY BEGIN IN 1933?

It did, and yet it did not, because for the true beginning we must go back to Dale's early publication in 1914 on 'The action of certain esters and ethers of choline', in which he made the fundamental distinction between two types of action of acetylcholine, a muscarine and a nicotine action: that some of the peripheral actions of acetylcholine resembled those of muscarine, others, those of nicotine. Muscarine actions were those on smooth muscles, heart muscle and gland cells, and they were sensitive to atropine; nicotine actions were those on sympathetic (and parasympathetic) ganglia and on the cells of the adrenal medulla, and they in turn were abolished by large paralysing doses of nicotine. Nothing was said at that time about the third nicotine action of acetylcholine: that on motor endplates of striated muscles. This action was not discovered until much later.

Dale's early paper showing the distinction between muscarine and nicotine actions of acetylcholine is the true beginning of the story of acetylcholine in synaptic and neuromuscular transmission. The paper is important historically also because of two predictions or glimpses into the future which later were turned into facts by work carried out in Dale's own laboratory.

Dale predicted by sixteen years the presence of a cholinesterase in blood when he pointed out that the extraordinary evanescence of the effects of acetylcholine injected intravenously was probably due to its hydrolysis into choline and acetic acid by an esterase in blood. The enzyme was discovered independently by Matthes in Dale's laboratory, and by Engelhardt and Loewi in Graz. In both publications which appeared in 1930, eserine was used to inhibit the enzymic destruction of acetylcholine. Yet the obvious conclusion was not drawn that eserine might provide a simple means for detecting acetylcholine released into the blood.

It was a glimpse into the future, his statement in 1914 that acetylcholine would be the ideal transmitter substance for parasympathetic nerve effects, if there were any evidence for its presence in animal tissue. This evidence was provided by Dale himself together with Dudley fifteen years later when they identified acetylcholine in extracts of horse spleen.

But we find no suggestion, not even a hint that the nicotine actions of acetylcholine, too, might have physiological functions. Such a possibility was not envisaged in 1914, nor later when acetylcholine was shown to be a natural constituent of animal tissue, and when the presence of a cholinesterase in blood was discovered. The idea did not even arise when the contractures produced in denervated mammalian striated muscles by arterial injections of acetylcholine were investigated, for instance, in 1930 by Dale and Gaddum in the denervated tongue, and a year later by von Euler and Gaddum in the denervated muscles of the upper lip. The aim of these experiments was to shed light on the contractures produced in these denervated muscles on stimulation of parasympathetic and sympathetic nerves, and to find an explanation for this strange behaviour of non-motor nerves in becoming 'pseudo-motor' as it was called, which had been known for so long. In 1863, the contracture produced in the denervated tongue on stimulation of the parasympathetic fibres in the lingual nerve was discovered by Philippeaux and Vulpian; it was confirmed by Heidenhain in 1883, and since then has often been referred to as the Vulpian–Heidenhain phenomenon. The contracture of the muscles of the upper lip on stimulation of the cervical sympathetic was described in 1885 by Rogowicz. The investigations of Dale, Gaddum and von Euler were a beautiful analysis and clarification of these pseudo-motor phenomena, suggesting that they were due to acetylcholine released on stimulation of nerve fibres belonging to the autonomic nervous system, but the thought did not arise that acetylcholine might be the neuromuscular transmitter. They did not even try their method of arterial injections of acetylcholine on the non-denervated muscle. Had they done so and had they injected acetylcholine in larger doses they would have observed real contractions. Would that have been the moment for the idea to be born of a physiological role of the nicotine action of acetylcholine? Or would they have described the effect just as a third nicotine action of acetylcholine, emphasizing perhaps, as I did a year later when I observed these contractions, that denervation produced two changes in the response to acetylcholine, not only rendering the muscle more sensitive to it but also converting the response from contractions into a contracture.

How near we all were just before 1933 to the idea of a physiological role for the nicotine actions of acetylcholine, and yet how remote from this advance. Today this may seem strange but, considering the ideas prevailing at that time about synaptic and neuromuscular transmission, it is not. We must remember that even the theory of chemical transmission from the endings of sympathetic and parasympathetic fibres to smooth muscles, heart muscle and gland cells was not yet universally accepted, although Loewi's experiments had gained wider and wider acceptance with each year. But the idea of chemical transmission occurring at the ganglionic synapse or at the neuromuscular junction was against the views of the electrophysiologists working in this field. They were convinced, and able to convince their colleagues, that with electrical methods alone it would be possible to unravel the mysteries of these transmissions, taking it for granted they could only be electrical, brought about by the eddy currents of the nerve impulse. Today we admire the ingenuity of their theories explaining these transmissions in electrical terms, the detonator response of Eccles, or the theory of Lapicque, that for transmission to occur the chronaxie of the nerve fibre must be the same as that of the innervated tissue.

Even after the first experimental evidence was obtained for the role of acetylcholine in these transmission processes, it took several years for the electrophysiologists reluctantly to give up their prejudice against a 'pharmacological interference' in these transmissions. Any result contradicting the acetylcholine theory was greeted with satisfaction. Lorente de No published experiments on the perfused superior cervical ganglion in which he found that acteylcholine appeared in the venous effluent on antidromic stimulation of the sympathetic fibres emerging from the ganglion and on stimulation of the vagus. If these results had been confirmed – but they were not (MacIntosh, 1938) – they would have provided decisive evidence against the role of acetylcholine in synaptic transmission. When Lorente de No's publication appeared, both Eccles and I were still in Australia. A colleague from Eccles's lab in Sydney came to see me in Melbourne. He said he had to give me, instead of greetings, the following message from Eccles: 'Acetylcholine is all wet!' I replied by telegram, 'Prefer wet acetylcholine to dry eddy currents'; signed, 'Anonymous'.

Many years later, at a meeting of the Physiological Society at University College London, after a communication about experiments which were meant to show that acetylcholine could not be the neuromuscular transmitter, G. L. Brown called from the Chair: 'Feldberg, you will have to fight with your back to the wall.' My reply was something like, 'I prefer this to any fighting without acetylcholine.'

When the acetylcholine theory became more firmly established, forgotten statements were dug up to show that there had been scientists in the past who had considered and favoured chemical transmission in these events. In 1937, Dale published a note in the *Proceedings* of the Society on 'Dubois-Reymond and chemical transmission' to illustrate that 60 years ago, in 1877, Dubois-Reymond had quite clearly suggested and favoured a chemical transmission of the excitation from motor nerve endings to striated muscle. Then there was the brilliant anticipation of this mode of action by T. R. Elliott, who had been the first to conceive the idea of chemical transmission in the autonomic nervous system. In his Sidney Ringer Memorial Lecture, he discussed this possibility also for striated muscles and the electric organ of fishes; he even made an attempt to extract the active substance. 'I have tried in vain to discover an active substance in the muscle plates of striped muscles. And Professor Herring was also disappointed when he examined for this purpose the electric organs of the skate which are exaggerated motor end plates . . . But it is hard to forgo the belief that such discoveries lie in the lap of the future.' That was written in 1914. When, in 1942, Fessard and I wrote our paper 'On the cholinergic nature of the nerves to the electric organ of the Torpedo', Dale drew my attention to Elliott's lecture and we quoted his prediction. I was so impressed by it that I was anxious to meet Elliott, and Dale, who knew him well, invited us both for dinner one evening at the Athenaeum. There I asked Elliott what kind of extraction they had used. As far as he could remember they had simply used water or saline extracts. If only they had used hydrochloric acid and thus prevented the spontaneous and enzymic hydrolysis of acetylcholine they could hardly have missed the high activity in the extracts from the electric organ. What a fantastic near-miss it had been!

HOW DID IT BEGIN IN 1933?

And how did it happen that I was in Dale's lab at that time? This, at least, I have no difficulty in answering.

One day in 1933, shortly after Hitler came to power, the Director of the Institute in Berlin where I was working sent for me and informed me that I had been dismissed, must leave the Institute at the latest by midnight that day, and was not allowed to enter it any more.

A few weeks later, someone told me that the representative of the Rockefeller Foundation was staying in Berlin and that I should try to see him. I succeeded. He was most sympathetic, but said something like this: 'You must understand, Feldberg, so many famous scientists have been dismissed whom we must help that it would not be fair to raise any hope of finding a position for a young person like you.' Then, more to comfort me, 'But at least let me take down your name. One never knows.' And when I spelt out my name for him, he hesitated, and said, 'I must have heard about you. Let me see.' Turning back the pages of his diary, he suddenly said, delighted himself: 'Here it is. I have a message for you from Sir Henry Dale whom I met in London about a fortnight ago. Sir Henry told me, if by chance I should meet Feldberg in Berlin, and if he has been dismissed, tell him I want him to come to London to work with me. So you are all right,' he said warmly. 'There is at least one person I needn't worry about any more.'

Here I must explain that I had known Dale since 1927, when I was a visiting scientist in his department. In 1930, Schild and I had dedicated a book to him on histamine, and in 1932, I met Dale again in Wiesbaden at a meeting of the German Pharmacological Society followed by the German Internisten Kongress where Dale was the invited lecturer. At the meeting of the Pharmacological Society (Otto Loewi was there, too) I gave a communication on the release of an acetylcholine-like substance from the tongue into the blood during stimulation of the chorda-lingual nerve, illustrating how easy it had become to detect the released acetylcholine once its destruction was prevented by an intravenous injection of eserine and the venous blood was allowed to pass over the eserinized leech muscle preparation. Another communication on the leech muscle preparation itself was given by Minz who at that time was working with me in the Physiological Institute of the University of Berlin. Dale was greatly interested in our communications.

Having received Dale's message from the representative of the Rockefeller Foundation, I wrote at once to Sir Henry, immediately received an invitation, and as soon as all necessary formalities were completed, I was literally thrown out by my wife who, with our two small children, stayed behind to get everything ready and packed for what might, and eventually did, become an emigration. I landed at Harwich on 7 July 1933.

This is how I happened to be with Dale in 1933, and worked in his laboratory for over two and a half years until we left for Australia which, as shown by my Certificate of Registration, was on 15 April 1936.

The first weeks in London were terrible, with the constant fear something might happen to the family in Germany and prevent them from leaving. When they finally arrived, I travelled to Harwich the night before to meet the boat due to land at about five the next morning. One could not help being in a state of anxiety because it was believed that passengers were not infrequently taken off the train at the Dutch frontier and sent back. Hours before the boat was due I walked up and down the empty quay, looking into the empty rooms until they filled with customs officials. One of the immigration officers must have observed me, because when he handed the landing permit to my wife, he asked her if it was me who had been waiting outside, and then said, 'Mrs Feldberg, you must never again leave your husband alone.' That was the compassion shown at that time by a large part of the English people to the refugees from Germany: that's what we had become.

But the following two and a half years were an unbelievably happy time.

Naturally one was deeply concerned with the fate of relations, friends, colleagues, and all those who were still in danger, and one tried to help whenever possible. But one could not help being happy; it was like a new life after a severe illness: to be safe with the family and in addition to have the possibility to do research under the most favourable conditions imaginable. Later, colleagues have often said to me that those years in Hampstead after 1933 must have been an exciting time. I, personally, have never been able to separate from each other our happiness and the work going on in the lab.

We had rented a two-roomed furnished flat about ten minutes' walk from the Institute in Hampstead, and nearly every morning my wife accompanied me to the lab, wished me good luck for the experiment and asked me not to worry however late it might be in the evening, just to ring her before leaving and she would meet me half way. One morning on the way to the lab we had stopped in front of a fishmonger's shop when Lady Dale passed, and asked why we were staring at lobsters. My wife explained that I was so fond of them, that whenever I felt we had done a really exciting experiment resulting in a definite advance, we would celebrate it with a lobster. When Sir Henry came into the lab next morning, the first thing he said was that, from then on, lobster experiments would be celebrated in his house. Sometimes when Sir Henry had to be away during the day at some meeting and could not be with the experiment, he would rush into the lab in the evening before dinner and ask, 'A lobster experiment today?' Synaptic and neuromuscular transmission have been responsible for a good number of lobster dinners!

Dale liked to start early in the morning, and realizing that I would most probably be late, I was often commanded to have breakfast with him at Mount Vernon House, to make certain the experiment would begin on time.

For the first six months I could support myself, then my accounts in Germany were frozen, but Dale obtained a Rockefeller grant for me for a year, which was paid quarterly. One morning, near the end of a month, Dale came into the lab whilst I was preparing our cat, and asked how often I still expected a quarterly payment. I said, 'once more'. Five minutes later, he came back and said, 'you were wrong, you have had your last payment; how long will you be able to go on?' I said, 'for another fortnight'. Collison, Dale's technician, who was assisting me, got really worried and asked what I was going to do. 'Let's go on with the experiment,' I said, 'I can't do anything; either Sir Henry succeeds in finding financial support, then all is well. Otherwise it is even more important that we do a few more good experiments in the time that is left.' Those were happy days. We had learned not to worry about trivial things. Dale naturally succeeded in getting the Rockefeller grant prolonged for another year, but he was furious with me that I had not been able nor willing to take the matter seriously.

Lady Dale told my wife how glad she was that we had come to England, because before then her husband had stopped working in the lab and now he was enjoying doing experiments again. The best proof that Dale had enjoyed the time I worked in his lab is the fact that before he retired in 1940, he told me that if it had not been for the war, he had hoped, and would have liked best, to work with me again after his retirement.

Dale's last experiment paper before 1933 was with Bauer, Poulsson and Richards, and on an entirely different subject, on liver perfusion. It appeared in the April number of the *Journal of Physiology* in 1932 and must therefore have been sent in some time in 1931. The one prior to this was the one with Gaddum in 1930 on acetylcholine contractions in denervated mammalian striated muscles discussed on page 66. After I left for Australia, Dale again stopped doing experiments.

Between 1933 and 1936, the index volume of the *Journal of Physiology* for volumes **61** to **100** lists, from Dale's laboratory, fourteen publications, printed communications or full papers, on acetylcholine in ganglionic and neuromuscular transmission, and another ten publications on transmission by acetylcholine in other tissues: in the suprarenal medulla, stomach wall, sweat glands, salivary glands and central nervous system. I cannot help it, but my name appears as co-author on all twenty-four of them. What is the explanation? And what was my contribution which resulted in this embarrassing fact, that I was an eye-witness in all of them?

To make use of a metaphor: perhaps it was that I had brought with me a key that would open the doors. Dale and Gaddum seemed to know what lay behind them, but I had the key. So I was asked to open first this one, then that one and so on, one after the other; and I never refused to do so.

And what was the 'key' that I had brought with me? It was my method of detecting acetylcholine released during nerve stimulation. It was the use of eserine, its intravenous injection or, in perfusion experiments, its addition to the perfusion fluid in order to inhibit the enzymic destruction of the released acetylcholine and then to detect it in the venous effluent, blood or perfusate, by means of a simple, specific and sensitive test, the eserinized leech muscle preparation. In the historical abstract to my demonstration on acetylcholine release from the perfused sympathetic ganglion, I have told the story of how it came about that this preparation became our routine test for assaying minute amounts of acetylcholine.

With this 'key' we obtained the first direct experimental evidence for the role of acetylcholine both in ganglionic and neuromuscular transmission. The second evidence followed in quick succession. It, too, was based on the effect of eserine to inhibit the enzymic destruction of acetylcholine. Eserine should affect the response to nerve stimulation, if due to acetylcholine, in a way consistent with persistence of undestroyed acetylcholine at the site of release. Eserine had been shown before to affect the response to nerve stimulation in this way but it had not been used so persistently for this purpose as we now used it, in order to obtain evidence for the role of acetylcholine in transmission processes. During these experiments it became evident that the effect need not necessarily be a potentiation of the response; eserine could have the opposite effect and produce attenuation due to a paralysing action of excess acetylcholine, for instance in the sympathetic ganglion.

With this two-fold experimental evidence the cholinergic nature of ganglionic and neuromuscular transmission was actually established although it took years before it was generally accepted. For the conversion of the strongest opponent, Eccles, we had to wait for over ten years. Those interested in the controversy during these intervening years cannot do better than to read the charming little book by Bacq on *Chemical Transmission of Nerve Impulses* which was presented to the participants of the Sir Henry Dale Centennial Symposium in Cambridge on 17–19 September 1975. Bacq dedicates a chapter to the controversy. He remarks that in 1943 both Eccles and I had a paper in the *Journal of Physiology* in which we discussed the problem of synaptic transmission. Eccles does not mention a single time the term acetylcholine, but states categorically that it has now been established that neuromuscular and synaptic transmission are mediated by the local negative potential set up by the nerve impulse. I concluded from experiments on acetylcholine synthesis that such synthesis is a property of the preganglionic endings in sympathetic ganglia and necessary for normal and particularly for sustained synaptic transmission. Bacq writes, 'In his work Feldberg ignored Eccles' arguments, and he did not even bother to refute them. The two groups knew one another, but there was a wall between them; each pursued his own research with his own basic concepts and methods. The ability to consciously exclude each other's views so rigidly was astounding, the more so in that the two papers appeared in the same issue of the *Journal of Physiology*.' Bacq was right. But why should I read Eccles? He did not believe in acetylcholine and we were convinced of its role in synaptic and neuromuscular transmission. As Taoism tells us, a good man does not argue!

The strong opposition of Eccles and others, however, exerted a most beneficial effect. We were not allowed to relax, but were forced to accumulate more and more detailed evidence in support of our theory.

Perhaps I might illustrate by another instance, the difficulties electrophysiologists had at that time, in switching over completely in their way of thinking from electrical to chemical transmission, even when accepting the transmitter role of acetylcholine. I referred to experiments on the electric organ of the *Torpedo* by Fessard from Paris, and myself. In 1939, he invited me to work with him at the maritime station in Archachon, in the South of France, so that we might apply to the electric organ the methods which had proved so successful in demonstrating the role of acetylcholine in neuromuscular transmission.

The electric organ of the *Torpedo*, as well as of other fish, may be regarded as a collection of modified and exaggerated motor endplates lacking the contractile structure of voluntary muscle fibres. In the *Torpedo*, the organ consists of a great number of prisms arranged side by side, each prism being built up of several hundred superimposed plates all of which are covered on their ventral side by a terminal nerve net. At the moment of the discharge the ventral sides of all plates become negative to the dorsal nerve-free sides; the discharge produced by a whole organ may amount to several hundred volts. Auger & Fessard (1938, 1939) had done some fundamental work on the isolated prisms and, from the latency and form of the discharge on electrical stimulation, had been able to conclude that it is produced indirectly and that the organ cannot be activated electrically by the impulses reaching the terminal network. In addition, there was the very suggestive fact that the organ was an extremely rich source of cholinesterase (Marnay, 1937; Nachmansohn & Lederer, 1939).

Fessard and I having found that the organ was also rich in acetylcholine and that on stimulation of its nerves acetylcholine appeared in the effluent when the organ was perfused through its artery with the appropriate salt solution, which naturally had to contain eserine, we also wanted to know if acetylcholine had the electrogenic property required to qualify as transmitter. So the perfused organ and I were imprisoned in a small screened cubicle whilst the recording was done in freedom outside. When all was ready for the recording, I was given the word 'go' and at once made my arterial injection of acetylcholine. It resulted in an angry shout from outside. I was told I had nearly wrecked their apparatus. I had to be insulated, had to put on rubber gloves, and when that did not help, rubber boots as well. But even that did not help, and the engineer in charge of the recording refused to go on, being afraid for his equipment. I begged him to allow me just one more injection, and when it was made it was greeted with a joyful shout. The engineer opened my prison and said beamingly: a perfect injection, and absolutely no effect. I had injected the salt solution without acetylcholine. So I told him what had been wrecking his apparatus was my acetylcholine. From the discussion which followed, in rapid French, unintelligible to me, between Fessard and him, it came out that the engineer did not believe in acetylcholine, but he had wanted to give us a fair chance and therefore had recorded with the highest sensitivity. When he reduced the amplification and allowed a fraction only of the discharge to enter the amplifier, and I reduced the dose of acetylcholine, its injection produced a beautiful deflexion, still too large to be recorded in its entirety, but no longer frightening to the engineer.

The real reason for referring again to our experiments on the electric organ, however, was a different one. When writing up the experiments, I suggested as title for our paper something like 'The cholinergic nature of the nerves to the electric organ'. But Fessard said, 'We cannot do this'. And when I asked why, he said, 'Because l'organ électrique n'existe pas'. I was dumbfounded and amused. Here we had been working for weeks on an organ which certainly weighed more than a pound, and then to be told it does not exist! But Fessard explained. 'You must realise,' he said, 'that the organ, once its nerves have degenerated, is inexcitable.' This was shown by Garten in 1910, and confirmed by Fessard himself when working on the isolated prisms. He showed that they discharge normally at the slightest provocation not only on electrical stimulation but also when lightly pressed upon, say, by the tip of a pencil. However, once the nerves to the organ have degenerated, it is impossible to obtain a discharge. The electric organ is thus but part of the nervous mechanism and does not exist on its own. This interpretation illustrates how easy it was at that time to fall back into thinking in terms of electrical transmission, because that is what it was. A relapse, but a momentary one, because it did not take us long to find the interpretation in terms of chemical transmission, that the sole physiological stimulus able to excite the electric organ and cause it to discharge is acetylcholine. Electrical and mechanical stimulation of the organ or its prisms does nothing but excite the nerve fibres or nerve nets at the ventral surface of the plates to release acetylcholine which then produces the discharge; once the nerves have degenerated there is no acetylcholine available to be released by whatever stimuli; therefore no discharge is possible. According to this interpretation arterial injections into the perfused organ should produce a discharge also after degeneration of its nerves. This we intended to show. Fessard was going to operate on some fish and a few months later when the nerves had degenerated, bring the survivors by tank to Paris. I, in the meantime, would try to obtain a grant enabling me to fly to Paris about once a month so that together we could perfuse the denervated organ. I made a bet with Fessard that acetylcholine would cause a discharge. A few days after my return to London, I saw Dale at a meeting of the Physiological Society, talking to the representative of the Rockefeller Foundation. When Dale saw me he waved me over, and I had to tell both of them our results. What a chance!

The outcome was a grant to enable me to fly over to Paris once a month. I remember Dale asking me if I would do so also if one fish only survived. I could not resist replying that I would do so even if none survived! The opportunity to make use of the grant, however, never arose. War started, and although I am certain that Fessard and I had the same opinion about the outcome of our bet, I have not been able to collect it – my bottle of champagne!

THE YEARS 1933 TO 1936

A few dates may illustrate the rapidity with which everything happened in 1933 and during the following months. The first publications from this period dealing with chemical transmission by acetylcholine were printed communications given to the meeting of the Physiological Society on 18 November 1933. As they had naturally to be sent in some time earlier, it meant I had not been working in Hampstead for much longer than three months.

One of the communications sent in was by Dale and myself on the vagal innervation to the stomach. The motor effects to the stomach, as those to the urinary bladder, were known to be resistant to the action of atropine. This had been an obstacle to the general acceptance of acetylcholine, or any choline ester, as the probable chemical transmitter of these parasympathetic effects. Dale's keenness to find out if such effects were also associated with release of acetylcholine was what started the experiments. In our communication we reported a manifold increase in the acetylcholine content of the venous blood collected from the stomach during vagal stimulation in dogs, provided naturally that they were treated with an intravenous injection of eserine.

Another communication was by Minz, Tsudzimura and myself on the role of acetylcholine as transmitter to the suprarenal medulla. Before I left Germany, Minz and I had found, with the help of intravenous eserine and of the eserinized leech muscle, that an acetylcholine-like substance appeared during stimulation of the splanchnic nerves in the venous blood from the suprarenals. We had continued this work with a Japanese colleague, Tsudzimura, when it was interrupted by my dismissal. But I was allowed in Dale's laboratory to continue and complete the work which provided the second piece of evidence for this transmission by acetylcholine: the potentiating effect of eserine on the pressor response produced by splanchnic stimulation to the suprarenals.

The third communication, by Gaddum and myself, was the one on synaptic transmission by acetylcholine in sympathetic ganglia. In my historical abstract to the demonstration of these experiments, I told the story of how it came about that we looked for acetylcholine in this transmission, but I shall tell it again. There were first two findings we had made independently of each other. The finding just mentioned, by Minz and myself, of the appearance of an acetylcholine-like substance in the blood from the suprarenals on splanchnic stimulation, that is, on stimulation of nerve fibres which correspond to preganglionic sympathetic neurones, and the pharmacological identification by Chang and Gaddum (1933) of acetylcholine in extracts of the sympathetic chain of the horse. They were much nearer the truth than we were when they suggested at that time that acetylcholine 'might play a part in normal transmission of impulses through ganglia'. How timid and vague this statement appears today! Then there was the additional strong impetus given by a publication of Kibjakow (1933) which had just appeared in *Pflüger's Archiv*, the description of a method for perfusing the superior cervical ganglion in a cat. Just what was needed! It is not so important today that his results have never been confirmed, that the effluent collected during stimulation of the cervical sympathetic when added to the fluid perfusing the same or another ganglion stimulated it. What was important and revolutionary at that time was his idea, and his attempt to test it, that a substance should be released from the preganglionic nerve endings and be responsible for the transmission of the nerve impulse across the synapse. Gaddum was fascinated and excited; he showed me Kibjakow's paper; I had to read it at once. But having done so, he had no difficulty in persuading me to join forces to use Kibjakow's method, but with the additional trick of adding eserine to the perfusion fluid, to test the effluent collected during stimulation of the cervical sympathetic on the eserinized leech muscle and, if it should cause contraction, to use the procedure Gaddum had so beautifully worked out with Chang, of identifying pharmacologically an active substance with acetylcholine by making parallel estimates on different assay preparations. And that was what we did.

But there was a fourth communication given by Dale at this meeting on 18 November 1933. It was the most important of all, partly the outcome of the findings reported in the other communications. It was his 'Nomenclature of fibres in the autonomic nervous system and their effects'. His introduction of the terms 'cholinergic' and 'adrenergic' fibres or neurones; the realization that the anatomical designation of sympathetic and parasympathetic fibres was not sufficient because it did not give any information of whether a nerve fibre acted through the release of acetylcholine or of 'something like adrenaline' which we now know to be noradrenaline. I remember well Dale having gone on several days to the Library, which was not usual for him, and then his happiness one day when he came to show me the typescript of his communication. He was fully aware of its importance.

With this nomenclature a new dimension was added to our understanding of the autonomic nervous system. It told us that probably all its preganglionic fibres, the sympathetic and the parasympathetic ones, are cholinergic and that there are exceptions to the general rule according to which the postganglionic fibres in the parasympathetic system are all cholinergic and those in the sympathetic system all adrenergic. One such exception mentioned in his communication was the innervation of the sweat glands, which in cats are found in the hairless pads of the feet. The secretory fibres to these glands are anatomically sympathetic, but as the sweating is abolished by atropine it appeared to be transmitted by something like acetylcholine. No wonder then that within weeks Dale and I perfused the cat's hind foot with Locke's solution containing eserine. And it came as no surprise to us that the sweating produced in the hairless pad on stimulation of the lower end of the abdominal sympathetic chain was associated with the appearance in the venous effluent of a substance which contracted the eserinized leech muscle and which had the other usual properties by which we recognized acetylcholine.

The nomenclature was given solely for fibres of the autonomic nervous system. No mention was made of the cholinergic nature of motor nerves to striated muscles. Yet again within months, Dale and I began perfusing with eserinized Locke's solution the cat's tongue, and found that stimulation of its motor nerve, the hypoglossus, with its sympathetic fibres degenerated, led to the appearance in the venous effluent of a substance which contracted the eserinized leech muscle and had the other properties of acetylcholine. We then obtained the same result on perfusion of the leg muscles of dogs on stimulation of the appropriate spinal ventral roots with the sympathetic fibres degenerated. These results provided the first experimental evidence for the cholinergic nature of neuromuscular transmission.

Certainly at the time when Dale wrote his printed communication none of us thought of this extension of cholinergic nerves to striated muscles. So when did the idea arise? I always thought it was some time after the meeting of the Society. But in *Adventures in Physiology* Dale comments on the communication and tells us that at the meeting he 'made a preliminary, verbal mention of our plan to test the significance of these findings, by analogy, for the transmission from motor nerve endings to voluntary muscle fibres'. With 'these findings' he referred to our experiments on the role of acetylcholine as transmitter from preganglionic sympathetic fibres to the ganglion cells and the medullary cells of the adrenals which he described as a 'rather daring excursion'. Yet for those not inhibited by the thinking of electrophysiologists at that time, it was nothing special to look for a physiological function of two pharmacological actions which had been known for nearly twenty years. On the other hand, it is difficult today to realize the immensity of the novelty of looking, as Dale expressed it 'by analogy' for the role of acetylcholine in neuromuscular transmission, because the mammalian muscles were usually regarded as being completely insensitive to the action of acetylcholine when their motor nerve supply was intact. The recognized pharmacological effect was the contracture after motor nerve degeneration.

I mentioned the rapidity in the development of the acetylcholine theory in 1933 and during the following months and that five communications from Dale's laboratory were given at the meeting of the Physiological Society in November 1933. Six months later, another five were given at the May meeting in 1934, also dealing with the transmitter function of acetylcholine. Two are of no interest to our problem. The third and fourth dealt with the experiments just mentioned on the cholinergic nature of the sympathetic innervation to the sweat glands of the cat and of the motor innervation to voluntary skeletal muscle.

The fifth communication dealt with the 'Action of eserine in transmission through the superior cervical ganglion'. It gave the second piece of evidence required for the acceptance of the role of acetylcholine in ganglionic transmission. It showed that the contractions of the nictitating membrane produced by preganglionic stimulation were potentiated when eserine was added to the fluid perfusing the ganglion in high dilution, but depressed when added in stronger concentration; and that eserine had the same effects on the contractions produced by small doses of acetylcholine injected into the perfusion. This communication was by Vartiainen, a visiting scientist from Finland, and myself. We could also give the first estimate of the number of molecules of acetylcholine released by a single preganglionic volley per synapse. Calculated from the amount of acetylcholine appearing in the venous effluent of the perfused ganglion, the estimate was about three million molecules.

All these printed communications came out shortly afterwards as full papers in the *Journal of Physiology* except the one on neuromuscular transmission, which took longer. The reason was that we felt we had to amplify our results to avoid unnecessary criticism because we knew the opposition to our theory would be strong. In these experiments we were fortunate in being joined by Marthe Vogt who had left Germany because she detested the Nazi regime and what it stood for. During the months after her arrival we extended the limited results so far obtained to other muscles, including the leg muscles of the frog and, in order to meet possible criticism that the muscular contractions were responsible for the release of acetylcholine, we showed that the release did not occur on direct stimulation of the muscles, provided their nerves had degenerated, but that it occurred after curarine which abolished the contractions. Curarine was thus shown to act on motor nerve stimulation like atropine on parasympathetic stimulation, preventing not the release of acetylcholine but its action. All these results were included in the full paper by Dale, Vogt and myself which was sent to the *Journal of Physiology* in January 1936. This paper, however, did not yet contain experiments on contractions produced by acetylcholine or on the effect of eserine on motor nerve stimulation; but they were under way, and published four months later.

Gaddum had left at the end of 1933 to take up an appointment as professor of pharmacology at the University of Cairo. I still find it difficult to realize how short this fruitful time was, less than half a year, that we had worked together. Dale must have felt that it would become essential at some stage in the not too distant future to apply electrophysiological methods to the theory of ganglionic and neuromuscular transmission by acetylcholine. So he offered the position which had become vacant to G. L. Brown. He accepted and started working in Hampstead in April 1934.

In the *Biographical Memoirs* of The Royal Society, MacIntosh and Paton say that Brown always liked using electrical apparatus and that his experience of working with Eccles's amplifiers and oscillograph confirmed his preference. He had worked with Eccles in Oxford for half a year, but before he came to Hampstead he was working with McSwiney in Leeds on the innervation of the stomach. 'I was so interested in how the impulse got there and what happened to the inhibited or excited muscle that the process of neuroeffector transmission left me cold,' wrote Brown long afterwards. What better choice could have been made!

With Brown's arrival electrophysiological methods were introduced into Dale's laboratory. During his first month there, Brown was preoccupied with building, almost single-handed, his own amplifier for recording action potentials with the aid of a Matthews oscillograph. Later he divided his time between improving his equipment, which to my recollection seemed to go on indefinitely, and doing experiments with me on the superior cervical sympathetic ganglion, mainly perfusion experiments, and later also on neuromuscular transmission. He preferred electrophysiological methods for analysis but was always keen to try others as well. The main results we obtained together during the following months on ganglionic transmission, I think, were the following six, which are given in chronological order:

(1) Potassium, as well as rubidium, released acetylcholine from the preganglionic endings of the perfused ganglion.

(2) The ganglion cells were not only stimulated by potassium (this had been shown previously by Vartiainen and myself) but also, with larger

doses, paralysed. The paralysing action on the cells did not affect the release of acetylcholine from the preganglionic endings.

(3) The ganglion became nearly depleted of its acetylcholine after degeneration of its preganglionic fibres.

(4) The action of curarine on the ganglion. It rendered the ganglion cells insensitive to acetylcholine, but not to potassium, and it did not prevent the release of acetylcholine from the preganglionic endings.

(5) Partial paralysis of the ganglion by excess released acetylcholine. In the presence of eserine the concentration of acetylcholine released in the perfused ganglion during the initial stages of stimulation was sufficiently high to do so.

(6) The release of acetylcholine during preganglionic stimulation was associated with synthesis of acetylcholine. During prolonged stimulation the amounts released were several times those obtainable from an unstimulated ganglion, yet its acetylcholine content did not diminish.

Another finding, though not a regular one, and the first of its kind concerning acetylcholine synthesis, was the ability of choline to augment the release of acetylcholine from the perfused ganglion when its output had gone down during prolonged stimulation.

Electrophysiological methods were not required to obtain these results, although some recordings of action potentials of ganglia were made and published, and the methods were used to study the effect of potassium injections on facilitation between two successive volleys, but nothing new was added to the problem of the role of acetylcholine in ganglionic transmission.

The superiority of the electrophysiological methods and the necessity to use them, however, became evident for the experiments on neuromuscular transmission. Before describing them, I shall mention in passing that one of the first indications for the role of acetylcholine in central transmission dates back to this period: the finding by Schriever and myself that after eserine injections, adrenaline and asphyxia caused the appearance of acetylcholine or its increase, in the cisternal cerebrospinal fluid of dogs.

The story of neuromuscular transmission by acetylcholine is an instance where a special physiological function of a substance was discovered before its corresponding pharmacological effect. Before 1934, acetylcholine contractions of normally innervated voluntary mammalian muscles were practically unknown. True, there were a few isolated observations of this kind, but little notice was taken of them and their significance was not recognized. However, once acetylcholine was found to appear in the effluent from a muscle perfused with eserinized salt solution during stimulation of its motor nerve, it became essential to 'discover' this action of acetylcholine, because how could acetylcholine play a role in neuromuscular transmission if it were unable to contract voluntary muscles. So we had to make a real effort to obtain this evidence.

Reading through the publication by Brown, Dale and myself, the memory comes back of the struggle we had, and how long it took us to obtain 'decent' contractions. In the course of our experiments we found that in muscles perfused with physiological salt solutions the conditions were never optimal because even in the earliest stages of perfusion an incipient oedema would dilute the injected acetylcholine solution. We realized we had to study not only the effect of acetylcholine under natural circulation of the muscle, but also to reduce the volume in which the acetylcholine had to be injected and to make the injections into the muscle artery as close to the muscle and as rapid as possible. The method of 'close arterial injection' as it came to be known was worked out to perfection later on by Brown when he used it for his classical electrophysiological analysis of the acetylcholine effect in different muscles. But we obtained the evidence needed, because we could show that acetylcholine produces muscular contractions.

In the beginning we recorded the contractions of the perfused gastrocnemius muscle of the cat mechanically with a tension lever on a smoked drum. I remember how excited we became when, in one experiment on close arterial injection of a relatively small dose of acetylcholine (20 μg) a rapid contraction developed with a tension of nearly 7 kg which was more than three times that produced by a maximal motor nerve twitch. Although the duration of the contraction was longer than that of a single maximal twitch it was certainly too short to be a contracture. The question which haunted us at the beginning was, could the response be a contracture? Today this may seem strange but at that time it was not, because only acetylcholine contractures were known. Therefore our excitement when, during experiments with natural circulation of the muscle, even smaller doses of acetylcholine produced contractions that on the smoked drum appeared to be as rapid, or nearly as rapid, as single maximal motor nerve twitches, and when on changing from the smoked drum to optical isometric myograms the contractions really proved to be of only slightly longer duration.

We could give the correct interpretation of our results, that the response consisted of a repetitive firing of the muscle fibres due to persistence of undestroyed acetylcholine at the motor end plates, and that any contraction produced by the acetylcholine injections whether it was stronger or weaker than a single twitch 'would have the nature rather, of a short, asynchronous tetanus'. In the proofs we were able to add that electrical records obtained in the meantime by G. L. Brown proved that our deduction was correct.

W. FELDBERG

This interpretation had become obvious after we had analysed the potentiating effect of eserine on motor nerve twitches. How well I remember the very first experiment, and the next one as well, on the effect of eserine in a spinal cat. My suggestion was to use maximal single motor nerve stimuli. Brown was against it. 'What do you expect?' he said, 'the twitch is already maximal. You should know there is an all-or-none law!' 'But,' I said, 'what does it matter; let's see what happens; in all previous experiments on the release of acetylcholine we used maximal stimuli.' 'All right!' he said. And then our joyful surprise when, after the eserine injection, the response recorded on the smoked drum began to creep up and increased with each impulse to reach, during the following two minutes, double its original size. When I dared to remark that perhaps the all-or-none law did not apply, Brown became angry and retorted, 'Don't be a fool! Wait!' He had at once grasped the meaning of the result and was angry at not having thought of it before. In the next experiment we recorded the action potentials and saw: eserine converted the single 'twitch' into a waning 'tetanus'. 'What do you think now, Feldberg?' asked Brown triumphantly. 'You cannot remain a fool for long when working with great people.' What else could I reply? Brown and I loved to tease and to be rude to each other without personal offence. But we were all happy, having obtained so beautifully this last piece of evidence which was necessary for the acceptance of the acetylcholine theory of neuromuscular transmission.

EPILOGUE

Events seen by two eye-witnesses look different. Everyone knows this. Further, no judge would accept, without the greatest reservation, the account of an eye-witness after forty years. Memory of recent events diminishes with age, but that is not so with events that have taken place many, many years ago. They may even become more vivid, but only some of them remain with us 'like pockets of snow in a general thaw'. In addition, memory often plays strange tricks. So my account of the events at Hampstead in the 1930s must be taken with all the reservations normally applicable to an eye-witness's account.

In his little book on *Chemical Transmission of Nerve Impulses*, Bacq raised the question of whether, without me, research on chemical transmission of nerve impulses might have taken a different course. In my view, the role of acetylcholine in ganglionic and neuromuscular transmission was due to be discovered in one laboratory or another. Most unlikely that I alone would have stumbled on it, except if, through results obtained in other laboratories my eyes would have been opened. Gaddum was nearer to the truth than I was. But I doubt that he would have started the ganglion perfusion experiments alone before leaving for Cairo, or later on under the less favourable conditions there for doing experiments, and with the new teaching responsibilities. In Cairo his main research interest shifted to histamine. Two printed communications, not followed by full papers, were all that appeared from him and his Egyptian colleagues on the role of acetylcholine in chemical transmission, and these publications did not add much to our problem.

On the other hand, it is most likely, in my view, that work in Dale's laboratory after 1933 would have brought to light the role of acetylcholine in the two transmission processes even without my presence. Dale knew of Gaddum's finding of acetylcholine in extracts of the sympathetic chain, of Kibjakow's ganglion perfusion, of my finding of the appearance of acetylcholine in the venous blood from the suprarenals during splanchnic stimulation, and of my successful use of eserine for the detection of acetylcholine released during nerve stimulation. If the possibility of a physiological role of the nicotine actions of acetylcholine lay dormant in the mind of any one person, that person would have been Dale. I could well imagine him asking the next visiting scientist to his lab to use Kibjakow's method of ganglion perfusion and later to perfuse a skeletal muscle with eserinized salt solution, thus getting the results we had obtained. On the other hand, one cannot be certain, because Dale had stopped doing experiments before 1933, and again after 1936. Dale loved to dwell on the role played by lucky accidents in his research. Perhaps my dismissal in Germany on account of Hitler was yet another of Dale's lucky accidents, certainly lucky for me.

People have often asked me if I was sad to leave Hampstead in 1936. Nearly every visiting scientist working in Dale's lab was sad to leave. In this sense I was sad, too, but not in any other way. I knew that for the further advance in our insight of the role of acetylcholine in chemical transmission, entirely different methods would be needed from those simple ones at my disposal, which had been so successful at the beginning of our story. I would no longer be able to contribute much, and I might have outstayed my usefulness. We left with the rare sense of happiness which comes through the knowledge of having been useful – in an exciting period of rapid scientific advance.

SOME MEMORIES OF VISUAL RESEARCH IN THE PAST 50 YEARS

W. A. H. RUSHTON

STRUCTURE

We always have to start with structure. All the books start with structure. You can't know how an organ works unless you know the structure of what is working. Yes, yes! But pages of anatomy are utterly indigestible unless one can appreciate what part the structure plays in the working of the organ. And to describe in detail what is there is so much easier than to discover what part it plays that the great chapters on minute anatomy – those deserts of detail without a living functional watercourse, only a mirage from unverified speculation – are nearly unreadable. That chapter is like a dictionary, not to be perused from cover to cover, but to be consulted, a word at a time to throw light on some particular obscurity.

As minute anatomy goes, however, that of the retina is the best dictionary in the whole of the central nervous system (CNS). This thin and easily detached neural sheet may yield its secrets to histological techniques with a minimum of distortion. And the orderliness of the layered structures gives to the retina a more coherent and integrated picture than other parts of the CNS.

Long before the beginning of our period, Cajal had worked out the structure of retinal organization nearly as fully as could be resolved by the light microscope. He used mainly the Golgi impregnation technique, the success of which results from the fact that only 1 or 2 % of the tightly packed nervous elements will 'take' the stain and those cells are stained throughout.

So the section lies transparent under the microscope except for a few lucky sample cells stained black and sharp throughout their whole ramifying structure. I suppose their gambling spirit keeps Golgi histologists at their microscopes like gamblers at Las Vages in the grip of their 'one-armed bandits'.

Over the past fifty years Cajal's work has been confirmed and extended by many, notably in the systematic study of Polyak (1941). Then came the electron microscope !

The trouble with enormous magnification is to know where in the (retinal) world we are, and what we are looking at. Cells, familiar by light, look quite different by electrons, so the experts have to tell us what's what. Elegant is the trick of W. Stell in which a cell is stained by the Golgi method and photographed, and then this cell (with surroundings) is fixed, sectioned and mounted for electron microscopy. The final photograph shows the cell defined by its silver precipitate set in the minute detail of the electron micrograph.

A most important feature in electron micrographs is the presence of droplets called 'vesicles' on one side of the synaptic membrane. It is generally held that nerve signals pass from that side across the membrane – in fact that the vesicles contain transmitter substance. To be able to define in this way the direction of nerve transmission is clearly of cardinal importance in understanding the direct and feedback circuits etc. in nerve organization.

VISUAL PIGMENTS

Kühne followed up Boll's observation that a frog's retina dissected in dim light is bleached in strong light, i.e. turned from pink to white. Already last century he had established the principal reactions of rhodopsin, the bleaching of visual purple first to visual yellow and then to visual white, and the subsequent regeneration in contact with the pigment epithelium, faster from yellow, slower from white. He had also seen that rhodopsin was contained in the outer segments of rods and was absent from the human fovea (of an executed criminal). Schultze had shown that vertebrates active in daylight had cone-rich retinas, night animals rod-rich. So rhodopsin in the rods should serve night vision. This was proved when König and Trendelenburg showed that two lights of different wavelength adjusted in intensity to look identical by twilight (rod) vision, bleached equally fast the rhodopsin of the frog's retina.

RHODOPSIN IN SOLUTION

Great improvement in the study of rhodopsin bleaching followed Katherine Tansley's technique of bringing rhodopsin into solution with clear digitonin (instead of Kühne's somewhat coloured bile salts). In this solution the bleaching products of rhodopsin (Kühne's visual yellow) was studied with spectroscopic accuracy by Lythgoe in England and Wald in U.S.A. in the late 1930s.

Wald established that rhodopsin and its breakdown products were carotenoids and he named the product obtained by extraction with petroleum ether 'retinene'. He did not discover its composition, but Morton showed it to be the aldehyde of vitamin A. Lythgoe showed the products from the bleaching of rhodopsin were first *transient orange* (which quickly turned into Kühne's visual yellow), the colour of which depended upon the pH of the solution, and hence Lythgoe named it *indicator yellow*. Wald later misnamed visual yellow 'retinene', though that is no indicator.

But Wald and Hubbard (his wife) went on to make the great visual pigment discovery of the century. It illustrates the dictum 'You don't really understand a thing until you can make it.' They made rhodopsin, synthesizing it from ingredients none of which came from the eye except 'opsin', the protein with which the carotenoid is combined.

The carotenoid itself was added in the form of codliver oil, and the rhodopsin formed could be bleached by light in the same way as natural rhodopsin.

In order to be sure that all ingredients for the synthesis were pure, they replaced the codliver oil by pure crystalline vitamin A. No synthesis occurred!

It is the experimental setbacks, the non-fulfilment of our confident expectations, that provoke great discoveries in those competent to make them. Hubbard and Wald discovered the photo-isomerization of retinal by light. Vitamin A aldehyde (Wald's retinene, now called retinal) is a molecule that can exist in several isomeric forms. It occurred to Hubbard and Wald that perhaps the crystalline form they had used and found ineffective was the wrong isomer. After some trouble they obtained crystals of all the principal isomers of retinal and found again that the all-*trans* which they had previously tried was ineffective, but the 11-*cis* isomer was converted into natural, bleachable rhodopsin.

I remember, still with a thrill, the occasion when Dr Ruth Hubbard demonstrated this synthesis to me in her Harvard lab. The straw-coloured solution of opsin, 11-*cis* retinal and other ingredients was left in the dark and a little later it had turned into a splendid ruby fluid. 'Is it bleachable?' I asked. Into the recording spectrophotometer went the solution. The pen initially drew the absorption spectrum of rhodopsin, and this gradually changed to indicator yellow as bleaching proceeded; the solution withdrawn at the end was indeed yellow.

The great importance of the isomeric configuration of retinal is that its change is precisely what light does to start the train of events from which vision results. The action of light is to change the retinal component of the rhodopsin compound from 11-*cis*, which is stable in its protein nest, into the ill-fitting all-*trans* isomer, a restless inmate that causes the molecule to undergo a series of thermal reactions (bleaching) that ultimately generate rod nerve signals.

Since after bleaching, retinal is left in the all-*trans* configuration, which is the form ineffective for rhodopsin synthesis, a re-isomerization is necessary for rhodopsin regeneration. Hubbard, and many others have shown that there are two ways in which the 11-*cis* isomer may be reformed: (i) in the dark by an isomerizing enzyme, and (ii) in very bright light by photo-isomerization. The catch of a quantum that may flip the retinal molecule from 11-*cis* to all-*trans* may equally (with about the same probability), if falling upon all-*trans*, flip it back to 11-*cis*. Regeneration by photo-isomerization is of special importance in the visual cycle of insects and other invertebrates living in bright sunshine.

Digitonin extracts of photo-pigments from the eyes of land vertebrates yield mainly rhodopsin, but Dartnall and colleagues have obtained several different pigments from fish eyes, carefully analysed by partial bleaching with lights of various wavelengths.

He found that their 'density spectrum' plotted on a scale of wave frequency ($= c$/wavelength) was always the same shape, but lay shifted along the scale to different frequency maxima. This relation he expressed for convenience as a nomogram.

PIGMENTS *in situ*

When visual pigments are studied in extract, i.e. in neat rectangular troughs containing the digitonin solution, the uniformity of the optical pathways makes the density readings easy to interpret. This condition, however, is very different from that of the pigment within the receptors in the retina, especially with regard to the kinetics of bleaching and regeneration. Indeed, in solution the only regeneration that commonly occurs is the photo-regeneration from intense flashes. Thus, if we ask 'How does visual performance depend upon the nature and state of the visual pigments in the living eye,' it is important, for our answer, to be able to measure at various wavelengths the kinetics of bleaching and regeneration in the living eye. This may be done by reflexion densitometry in man.

REFLEXION DENSITOMETRY (RUSHTON, WEALE, ALPERN)

With this instrument light is sent into the eye and reflected back from behind the retina as in an ophthalmoscope. But the reflected light, instead of entering the eye of the physician is received by a photocell. This light has been twice through the retina and suffered absorption by the visual pigments there; consequently, the denser the pigment present the less the photocell response. In this way the density of visual pigment may be measured in man at any wavelength throughout any bleaching-regeneration manoeuvre.

If the retinal area from which the photocell light is collected is the rod-rich periphery, the pigment measured is almost entirely rhodopsin. If the light returns only from the fovea, the pigment is almost entirely from red- and green-sensitive cones (the fovea being blue-blind). These cone pigment measurements have contributed to our knowledge of colour vision and colour defect (see later).

DARK ADAPTATION

The pyramid builders must have been familiar with the fact that on going from the bright Egyptian sunlight into the dark passages of the tomb, perhaps to paint on the walls, at first hardly anything can be seen and even torches seem to cast little light. In fact, it takes about half an hour in the dark to regain the eye's full sensitivity. Kohlrausch in Germany and Hecht in New York made quantitative studies in the early 1920s of the recovery of sensitivity in the dark following strong light adaptation. They plotted against time in the dark the logarithm of the least intensity of a light flash that could be detected at that moment. The curve so plotted is called 'the Dark Adaptation curve'. This curve is easy to obtain and thousands of good curves have been plotted in research projects and classrooms. Both Kohlrausch and Hecht found that the recovery curve exhibited two branches: first the cones and then the rods. The cones recover faster and thus, at first, have the lower threshold, and they have fully recovered in about five minutes. But the slow rods continue to improve their sensitivity very greatly, and after a time – perhaps ten minutes – rods become more sensitive than cones and continue to improve for a further twenty minutes. Hecht proved by many clear experiments that the early branch represented cone excitability, the late branch rods. And this being established, he was able to use these two branches to study separately some rod and cone properties.

Why is it that bleaching away some of the rhodopsin raises the visual threshold of the rods? Wald made the plausible suggestion that when rhodopsin was removed by bleaching it was no longer there to catch quanta. When 50 % had been bleached then, the light intensity must be doubled for the same quantum catch. This turns out to be an inadequate guess: a 50 % bleach does not double the threshold, it increases it a million-fold.

We still do not know why the bleaching of the visual pigment in rods or cones raises so enormously the visual threshold of those receptors. But reflexion densitometry allows us to measure what is the level of visual pigment at each stage of dark regeneration and so to correlate it with the measured log threshold in the dark adaptation curve at that stage.

The experimental result is simple, though its explanation is not. It is found that bleaching a visual pigment raises the *log* threshold of the receptors containing that pigment by an amount proportional to the fraction of pigment bleached. And in the subsequent dark adaptation the same relation holds; as the pigment regenerates, the log threshold falls, remaining above the fully dark level by an amount proportional to the fraction of pigment still in the bleached state. The constant of proportion is very different, as between rods and cones and in other animals it is different again (e.g. in Dowling's pioneer work where rod sensitivity was measured in rats by the *b*-wave threshold for the electroretinogram).

In the mid 1920s Hecht, at Columbia University, pushed his way to the front with immense vigour, good experiments and an objective chemical theory that claimed the universality and simplicity which we all love so well.

He worked first upon the reaction times of invertebrates and studied how this was related to the state of light and dark adaptation. He knew from Kühne and others that the rhodopsin level in the retina also depended upon the state of light/dark adaptation and that rods were excited by the 'bleaching' of rhodopsin. Thus was born the Photochemical Theory whose basis was as follows: (*a*) The level of rhodopsin depends upon the balanced reaction between the bleaching by the adapting light and the dark regeneration that occurs simultaneously and independently. (*b*) Various features of rod performance depend upon the rate of bleaching of rhodopsin by the stimulating light. This in turn depends partly on the stimulus intensity and partly on the rhodopsin level under adaptation.

The strength of this very plausible theory was its definiteness. The kinetics could be expressed mathematically, and the quantitative expectations could be checked against experimental performance. The theory here stated for rods could be applied to cones where obviously not rhodopsin but different cone pigments must be operative, with different kinetic constants.

Hecht and his colleagues performed good experiments on many kinds of visual performance such as visual acuity, intensity discrimination, flicker fusion frequency etc. measured as a function of adaptation level. They obtained excellent agreement with theoretical expectation. Since the theory predicts only the rate of bleaching, some subsidiary hypothesis was needed to relate this to performance, e.g. in visual acuity that the bleaching rate at two neighbouring points must differ by a fixed amount for two-point discrimination. This impressive body of coherent work is summarized in Hecht's article in *Physiological Reviews*, **17** (1937). It dominated thought on vision in the U.S.A. (especially for psychologists), and even now the observed changes in visual sensitivity with changes in light level are often called 'bleaching and regeneration', though these in fact were never measured.

When one looks at Hecht's review to see his treatment of dark adaptation (i.e. the slow recovery of sensitivity after exposure to a strong bleaching light), one is surprised. This phenomenon from which the whole photochemical theory started and which seems to be so simple to explain, is not mentioned at all! A grave difficulty had arisen. If the rod dark adaptation curve is the expression of the course of rhodopsin regeneration, it must take about half an hour to reach completion. It cannot therefore, underly the Weber–Fechner relation, the rise in threshold for a flash if it is presented upon a luminous background. For here, not after half an hour, but within a minute or so of the change in background, the new equilibria are established.

Hecht could not make his photochemical theory explain the slow dark adaptation curve without rejecting it as explanation for his splendid series of experiments on flicker, acuity etc. as a function of the luminance level. He was unable to bring himself to disinherit these fine children; so the photochemical explanation of dark adaptation had to go.

The end of the photochemical theory was initiated, however, by Hecht himself. At the beginning of the war (1939) Pirenne came to Hecht from the Netherlands and brought ideas that were better developed there than elsewhere. Max Planck had proved that light in its interaction with matter had a quantum structure and several Dutch scientists (Zwaardemaker, de Vries, Van de Velden and Bouman) had concluded that the weakest light we can see in optimal conditions must involve the absorption of only a few quanta in the rods.

Hecht, Shlaer and Pirenne's classic experiment involved two estimates of the minimum number of quanta that must be absorbed by rhodopsin to result in visible detection of the flash. (*a*) In optimum conditions for light detection, they measured the light energy falling upon the eye and passing through the pupil. To this figure, corrections were applied for transmission loss through the eye media (some 50 %) and for the fraction that passes through the retina without being absorbed by rhodopsin. This gave the result that only some 5 quanta absorbed were needed for detection. (*b*) Since quantum events are strictly random, a light flash that on average will result in 5 quanta being absorbed will sometimes result in more, sometimes less, than 5 with a frequency that may be predicted from statistical theory. Thus if the criterion for seeing a flash is that at least n quanta must be absorbed, the *frequency* of seeing can be predicted for various n values when flashes of various average quantum values are delivered. These expectations were then compared with the actual frequencies found by experiment. The results were found to correspond to an n-value of about 5.

These two very different approaches gave fairly concordant results. But the actual number of absorbed quanta needed depends upon the standard of reliability adopted by the subject in his detection – how readily he will blush if he says he has seen the flash when in fact it was blocked and no flash was actually presented (false positive). My friend Maurice Pirenne (the Belgian) blushes easily, my friend Maarten Bouman (the Dutchman) I suppose to be incapable of blushing. Pirenne needs 6 quanta to save his face, but Bouman will brave it out with only 2, as you may read in their papers.

All are agreed that when 5 quanta fall at random on a population of some 10000 rods, there is but a small chance that two quanta will hit the same rod. Thus if a double hit was the criterion for seeing we should very rarely see the flash in that experiment where in fact 5 quanta absorbed (on average) in the whole area were seen on 50 % of occasions.

We are forced to an astonishing conclusion. Rods have attained such perfection of sensitivity that they will respond to the catch of a single quantum by a signal that can spread through the retina far enough to combine with signals from four other rods to reach a magnitude that can be detected by the brain.

Hecht must have seen that his new discovery – that visual excitation was an affair of single molecules – was utterly removed from the mass action kinetics of his photochemical theory, and some drastic reconciliation must be made. Unfortunately he did not live to make it.

NOISE

When Hecht sent me a reprint of his (1942) paper with Shlaer and Pirenne, I wrote expressing my indignation at the disgraceful part played by the visual nerves. I was working at that time on motor nerves in the frog and was impressed by their splendid reliability, every impulse being conducted over the whole motor unit and exciting every muscle there. Rods had achieved the perfection of sensitivity, for they responded to the catch of a single quantum, and then, I was shocked to learn, they were let down by nerves so feeble or untrustworthy that it needed five rods to reiterate the message before it got through to consciousness. I do not remember exactly Hecht's answer, but it drew my attention to the importance of signal/noise considerations.

The brain is an important executive, like a Minister of State, too busy to see casual visitors. But he will see a deputation, for that is likely to have something more significant to say. The visual minister will 'see' a deputation of about 5 quanta if they all come together.

Two factors since then have brought signal/noise consideration into prominence. Perhaps the most immediate for physiologists is the great improvement in their amplifiers. If the spontaneous fluctuations in nerve records are nervous and not instrumental, the question arises 'How can the brain distinguish signals from noise if our reliable amplifier cannot do so?'

The other factor results from problems of long-distance telephone communications, especially in the U.S.A. Shannon and many others have developed the important subject of Information Theory, which is rather similar to thermodynamics.

Fechner already considered that the eye's absolute threshold was limited by the *eigengrau* or intrinsic light, which may be called 'receptor noise'. Rose made an important comparison between the eye and an 'ideal photoreceptor', and Barlow has developed the concept of receptor efficiency in studying the ratio of actual to ideal performance. The importance of noise as a limit to useful sensory input is now well recognized. And a sherry party often confirms the truth of the statement 'One man's signal may be another man's noise.'

ELECTRICAL RECORDING

The enormous proliferation of electrophysiological records during the past fifty years is due to the great advance in the technology of amplifiers and oscillographs. At first commercial amplifiers were not adapted to low voltage inputs so an important achievement was B. H. C. Matthews's differential amplifier with the steel-tongue mirror oscillograph. In addition to its use in his own researches, this instrument gave most of the records in E. D. Adrian's comprehensive survey of the electrophysiology of sense organs and the CNS. I remember an outstanding demonstration by Adrian and Matthews given to about 400 members of the Physiological Society in the large Cambridge lecture theatre in the mid-1930s. This was the dawn of electroencephalographic recording, our first view of 'brain waves', and we saw the waves being generated from the brain of Adrian.

A projection from an epidiascope onto the large screen of the lecture room, showed an ink-writer recording the electrical waves, as they appeared in Adrian's brain. We saw a rare moment when, with eyes closed, that active organ was voluntarily brought into relative quiescence, and the idling crowd of brain cells broke into a senseless, synchronous and repeated chant, as it might be 'Great is Diana of the Ephesians' – the α rhythm. As soon as the eyes were opened or the subject did mental arithmetic, each cell seemed to go about its own business, and synchrony was replaced by utter irregularity in the summation of highly individual responses.

But long before this in the 20s, Adrian and Rachael Matthews had analysed the responses from the optic nerve of the conger eel and demonstrated some aspects of neural interaction in the retina, e.g. the dependence of latency of response upon the area of retina illuminated, or whether one or four light spots were presented.

About this time Hartline discovered the experimental potentialities of the eye of the large primitive arthropod *Limulus*, and analysed them with consumate skill both of hand and mind. This work has always held great appeal for physiologists, for these nerves, straight from the photoreceptor cells, encode the light into spike frequency in just the way that we feel it should be encoded. By the mid-30s, Hartline had recorded from single optic nerve fibres in the frog. The delightful simplicity had all vanished in the records from these third-order cells, and the individual nerves differed greatly one from another in the quality of their responses. The majority of nerves, even, gave about the same response when light went on or off, which seemed a poor foundation upon which to build the perception of a luminous picture. In the vertebrate eye there are far more rods and cones than optic nerve fibres, so receptors cannot each have a private line to the brain. The information from very many receptors must be compressed into a message in one nerve fibre. And this message is in a code of crippling simplicity, not an information-rich pattern of dashes and dots, but just a nearly regular time sequence of unit spikes. What each optic nerve fibre signals is something like this: 'Fibre n speaking, now stronger, now weaker, now off.' But when it is fibre n that speaks, that in itself may signify, 'At such-and-such a retinal location a shadow is moving in the tempero-nasal direction,' or some other specific feature of useful integrated information. Evidently the simple outputs from a group of photoreceptors are processed in the layers of the retina, and each optic nerve fibre encodes a particular unit of information. No wonder that Hartline found that the *Limulus*-like simplicity between light and nerve response had vanished from his frog records.

THE ELECTRORETINOGRAM (E.R.G.)

This large, rather slow electrical response of the eye to light is very easy to record and very hard to analyse. The subject therefore constitutes an enormous bog in which some fine things float, but in which I shall sink if I venture far from the edge.

Whenever an investigator studies complex records for a long time, bumps and other features begin to declare themselves first as distinct personalities, and later as old friends. So he will recognize them like his friends in a crowd but cannot say enough about them for *me* to recognize them without more intercourse with them than I am generally disposed to give. And I have never seen why the separate features, though representing complex and interacting processes, are generally supposed to contribute to the record by simple addition. Most of the distinguished workers in the electricity of vision have used the E.R.G., and Granit's early qualitative analysis into three processes P_I, P_{II}, and P_{III}, has best stood the test of time and is still widely employed. The most conspicuous feature, the large *b*-wave, appears correlated with visual sensitivity. Karpe and Tansley some twenty-five years ago measured, during human dark adaptation, the thresholds for perception and for a *b*-wave estimated from calibration to be of fixed amplitude and found them to follow the same curve. And Dowling found that in the rat the rise in log threshold for the *b*-wave was proportional to the amount of rhodopsin bleached (measured by extraction), the same relation to bleaching as had log visual threshold in the human dark adaptation curve, though with a very different constant of proportion.

All this (and much more) does not mean that the *b*-wave records the electrical activity of the 'sensorium', but merely that adaptation, which scales down the response signal, does so equally for signals going to the 'sensorium' and to whatever generates the *b*-wave of the E.R.G.

What does generate this wave? Dowling and Miller have given convincing proof that it arises from the Müller cells, i.e. the glial cells that stretch radially right across the retina and (according to Walls's *Vertebrate Eye*) 'fill like glue all the spaces between the other cells'. All the same, the E.R.G. is a valuable measure of retinal organization and used with care, it can give information not easily obtained in any other way (e.g. Rodieck and Ford).

Granit was also pioneer in recording from single large ganglion cells in the retina. After a first paper with Svaetichin where extremely fine electrodes were used, Granit employed a platinum wire barely protruding from the fused end of a glass capillary. This rounded tip could be moved over the retina without tearing it and thus could be shifted close to single large cells. Their response pattern in the cat was similar to that found by Hartline in the frog (predominantly on/off responses). At low luminance the spectral sensitivity corresponded to that of rhodopsin (to porphyropsin in the freshwater tench); at higher levels cone responses were recorded from the *same* ganglion cell.

No cells were clearly fed by one type of cone only, nor rods together with only one type of cone. The mixed cone response was of two kinds, which Granit called 'dominator' and 'modulator'. The spectral sensitivity of *dominators* to monochromatic flashes of various wavelengths showed a broad curve similar to the photopic visibility curve. Probably all types of cone contributed to it. *Modulator* curves were narrow and showed a much sharper spectral response than any known visual pigment. If cones of different colours inhibit each other, then spectral regions, where their absorptions overlap, will be regions of mutual inhibition and thus could account for the observed modulator sharpening.

RECEPTIVE FIELDS

Hartline defined the *receptive field* of an optic nerve fibre as that region of retina where a light (or safer, a dark shadow) must fall to affect the nerve. It is obvious that the organization of the receptive field must be the key to the kind of information encoded by the nerve, and a great deal of study has been given to field organization. Barlow, in the frog, and Kuffler, in the cat, independently found that light in the centre of the field and light at the periphery inhibited each other. A flash that was uniformly distributed over the whole field, therefore, had little effect. Robson, Enroth-Cugell, Cleland, Levick, Rodieck, Fukada and very many others are among those who have analysed the different kinds of organization in receptive fields. Too much of this seems to be like a neurologist collecting 'signs' (the Babinski response for instance) as aids to distinguish one condition from the next, but with receptive fields there seems little attempt to find what is the essential feature of neurological organization that is responsible for the appearance of the sign. Those who offer new signs, and give new labels to the ganglia which respond to them, should follow Barlow and Levick in their ingenious analysis of movement detection. When the receptive field organization is understood sufficiently to explain the sign, then, and hardly before, will the sign have physiological meaning.

CORTICAL RECEPTIVE FIELDS

Receptive fields may be mapped on the retina not only from the responses elicited in single optic nerves, but from nerve cells further up the optic paths, e.g. lateral geniculate or striate cortex. The beautiful and extensive studies of Hubel and Wiesel have shown that cortical cells respond cniefly to lines and edges lying on the retina in a fixed direction, and to movement in the direction perpendicular. And the columns of these cortical cells perpendicular to the brain surface are clusters all with the *same* directional sensitivity.

Closely related to this appears to be the grating spatial analysis in the retina studied by Campbell, Robson, King-Smith, Kulikowski and very many others. They have found evidence for nerve channels to the brain specifically sensitive to the spatial frequency and direction of gratings, and relatively independent of each other. Contrast thresholds for gratings of various spatial wave forms can be predicted from Fourier analysis in the same way as de Lange's well-known prediction of the results of temporal analysis (flicker). Very many cortical cells are supplied by messages from the two eyes. It is upon the organization of these that our stereoscopic vision depends. This has been widely studied by Bishop, Henry, and their Australian colleagues. Since the time of Wheatstone's stereoscope it has been realized that the images of the visual scene formed on the two retinas will show small disparities depending upon the distance away of the object imaged. And this disparity is somehow interpreted by the brain into 'distance away' without loss of image sharpness. The nature of this interpretation is revealed by finding cortical cells that are specifically reinforced by inputs from both eyes when the disparity corresponds to an object at a particular distance, but inhibited by other disparities.

The discovery of cells in the visual cortex that respond to light in one or other or both eyes prompts the question, 'Is this organization inherent in the genetic make-up, or does it develop from visual input?' Important work upon this plasticity of organization started with Hubel and Wiesel's well-known analysis of the connexions of single cortical cells. They found for instances that when one eye of a kitten was shut by suture from birth, this eye made no connexion with the cortex, but when *both* eyes were sutured neither eye was so seriously cut off. Barlow, Blakemore, Mitchell and others have explored in many ways the plasticity of the visual organization, to see how performance depends upon the visual input in early life.

I shall end this section by mentioning a memory of G. S. Brindley, whose *Physiology of the Retina and Visual Pathway* is one of the Society's most distinguished Monographs. Brindley had already observed upon himself the effect of electric currents upon vision, by thrusting beneath his eyelids electrodes that could be apposed to the eyeball in various known positions. He observed the resulting electric phosphenes and inferred the structures excited by the current. At another time he measured the slow beats between flickering light and alternating currents passed through the eye above fusion frequency, to find where fusion occurs.

Now he wished to experience the effect of sending a strong current through the scalp to stimulate the brain. Strong currents are painful, and the strongest that even Brindley could stand only spread to the eye and generated what he proved was the electric phosphene there. He decided that he must anaesthetize the scalp and apply a much stronger current within an earthed enclosure.

I entered his room to ask something. There were two men in coats whiter than was common in our lab. They were physicians from the hospital who had just shaved and anaesthetized Brindley's scalp and driven a palisade of small nails into the skull. This fence was all connected by an earthed wire and was slowly bleeding. Behind the door was a lady holding a baby!! This was unusual even for a Brindley experiment. I blinked and retired. I think that experiment was inconclusive, but there was no scandal about the baby. Brindley was then press editor of the *Journal of Physiology*, and the lady was his assistant. She had come in with her baby, and a bunch of manuscripts, to ask some point.

A few years back I attended a conference on visual prosthesis. Brindley was not at the conference, but nothing reported at that meeting was so promising and exciting (I thought) as Brindley's experiments with the courageous blind nurse, his colleague and subject. She, understanding fully the implications of the operation, allowed her visual cortex to be fitted with a cap studded with stimulating electrodes each of which could excite a luminous phosphene 'about the size of a grain of rice' at a distinct point in the visual field. Brindley's achievement is still a long way from useful electric seeing, but he seems to be advancing well along that road.

RECORDING FROM PHOTORECEPTORS

A particularly important paper was that of Fuortes and Hodgkin analysing intracellular records from the eye of *Limulus*. They have a model – that's nothing, we all have a model that will explain pretty well our records, or at least those records we intend to publish – but what is so helpful is the way that they reveal how they came to choose that model.

The common practice with model salesmen is to specify, as in a sales catalogue, the features of their model and to proceed as though no other model need be considered. The justification of that model is that it mimics the actual performance of the organ over a certain range. The questions that we should like to ask (but usually we are too tactful to do so) are 'What other kinds of model would mimic performance just as well as yours does?' and 'Over what range of conditions does the mimicry hold?'

The interesting approach of Fuortes and Hodgkin is that they do not claim to have received from on high the specification of the *true* model. They ask 'What kind of a model on receiving an instantaneous light pulse will respond, not with a sudden step of electric output, nor yet with a sudden finite rate of rise, but by something a good deal more gradual?'

They take various features of the electric response to light, and ask what must be the nature of the model that will mimic this.

The reader feels that he is being taken into the authors' workshop and taught how to build a good model, which, however, is only provisional and capable of modification to meet new requirements.

With *Limulus* the model was a ten-stage, capacity-shunted amplifier with potential V arising from a light pulse $I(\Delta t)$ according to

$$V/I(\Delta t) = Ka^n t^{n-1} e^{-\beta t} \quad (n = 10). \tag{1}$$

Baylor and Hodgkin later, experimenting with single turtle cones, found the same formula with $n = 7$. Hodgkin has been continuing this analysis with Lamb and Simon on cells in vertebrate eyes.

The goal of minute retinal recording has long been to obtain intracellular records from single vertebrate cones, and, as usual, the Japanese have proved to be the most skilful. Tomita, who has already been preeminent in very many studies with minute manipulation, obtained single fish cone responses with Kaneko, Murakami and Pautler. The spectral sensitivity found was of three kinds and corresponded to the microspectrophotometric studies of Marks and MacNichol with the pigments in similar fish cones.

With invertebrate receptors (*Limulus*) light is found to depolarize and hence generate a current in the direction to stimulate the receptor nerve. In the vertebrate eye the effect of light is the other way round, as shown by Tomita, Toyoda and Baylor and Fuortes. Vertebrates release their stimulating transmitter substance continually in the dark (Trifonov) and the effect of light is to turn off the transmitter release and the dark current of Penn, Hagins and Yoshikami (rat)

One of the great difficulties in this minute recording has always been to know from what cell the records were obtained. A most beautiful technique used by Kaneko, and now by many others, is to fill the recording micropipette with a fluorescent dye (procyan yellow). After the record has been obtained, the dye is injected electrophoretically into that cell. When observed under the microscope in ultraviolet light that cell alone shines forth with a fluorescent light and great detail of cell form can be seen, and photographed.

The passage of signals across the retina from receptors to optic nerves has been studied by Werblin and Dowling in the amphibian *Necturus*, whose cells are large enough to be reliably penetrated by microelectrodes. They confirm that the familiar all-or-none spikes of nerve fibre conduction do not occur until the optic nerve fibres are reached. This limited code of unit spikes is reliable in conduction over long distances and hence important for the optic nerve, but within the layers of the retina information-rich signals of graded size are found, the signal being a variation up or down from a steady potential level.

Since Helmholtz, vision has been analysed by observing the regularities of psycho-physical relations and inventing the simplest model that will explain them. But the eye is not simple, and now we have access to structural and functional complexities that were hidden from Helmholtz. Psycho-physical relations are still what we wish to explain but the model will have to correspond to what is told us by the microscope and the micro-electrode. Today there is no scarcity of detailed observation. It pours in. The difficulty is to see how it can all hang together.

COLOUR

Long ago Granit warned me 'Colour is the *femme fatale* of vision. When once seduced, you will never be a free man again.' I was indeed seduced, and the fragments I here write upon this highly controversial subject are mainly some memories of my involvement.

Fifty years ago physiological students were taught that there were two rival theories of colour vision, Young–Helmholtz and Hering. In Cambridge I was at Emmanuel College and Thomas Young looked down upon me nightly from his portrait in the dining hall. So my loyalty was stirred, and I believed that there were three kinds of cone, red-, green-, and blue-sensitive, independently excited by light, and that colour sensation was determined by the proportions of those excitations. If this view were correct, then any colour could be matched by a suitable mixture of red and green and blue primary lights. This was first established experimentally with a colour top by Clerk Maxwell when he was a Trinity College research student and my loyalty to the Young–Maxwell theory was intensified when, after I became a Trinity Fellow, I saw on the parlour wall a photo of Maxwell in his early twenties, holding the top in his hand.

W. D. Wright, as soon as he had graduated at Imperial College, London, set about measuring accurately what mixture of red plus green plus blue primaries are required to match each spectral colour. He designed and built the most elegant trichromator in the world, and with it made colour matches of unsurpassed accuracy on normal and abnormal eyes.

When, 30 years later, Stiles made the measurements fresh for the International Committee (CIE) with many refinements, he confirmed Wright's results in nearly every particular.

If we knew exactly the spectral sensitivity of each of Young's three cones, we could deduce all the Wright colour-matching functions, so it might be hoped that the inverse would be possible, and that the three spectral sensitivity curves could be derived from Wright's well-determined mixture proportions. Unfortunately it is much easier to mix than to unmix and there is an infinite variety of possible cone spectral sensitivities that are mathematically equivalent as an exact basis for the Wright mixture results. So a good deal of endeavour during the past fifty years has been directed to obtain extra information to settle which spectral sensitivity curves are correct.

STILES'S TWO-COLOUR THRESHOLDS

It has long been known that a threshold flash ΔI must be made stronger if it is still to be seen when superposed upon a steady background I, and in most conditions Fechner's relation holds.

$$\Delta I = K(I + I_D), \tag{2}$$

where K is constant and I_D is his *Eigengrau* or 'retinal noise'. Stiles, in a research of great accuracy extending over twenty years, set himself to answer the question, 'What is the relation when the flash is one colour and the background another colour?'

When the intensities are all so weak that only rods are excited, the answer is clear and simple. All lights act on rods by being absorbed by rhodopsin. Any two lights of different wavelengths adjusted in intensity so that they are equally absorbed by rhodopsin will have identical exciting effects (*Principle of Univariance*). Thus flashes and backgrounds no matter what their wavelengths have only to be expressed in units of rhodopsin absorption (known from the rhodopsin absorption curve) and the wavelengths actually used become irrelevant, and we are concerned simply with the magnitudes of 'quantum catch' which obey the Fechner relation, equation (2).

When the intensities lie above cone thresholds a more complex pattern emerges. To a first approximation, we may say for rods and each of the three kinds of cone that the receptor is excited to the extent that its visual pigment catches quanta from the test flash and has its threshold raised to the extent that its visual pigment catches quanta from the background. Each receptor is affected by the quantum catch in its own type of pigment and is indifferent to the quantum catch elsewhere.

If this approximation were accurate it would lead at once to the spectral sensitivity (i.e. the quantum catch) of each kind of cone, but unfortunately the great extent and accuracy of Stiles's investigations has revealed many situations in which the foregoing generalization is not quite true.

KÖNIG'S CONJECTURE

In Berlin at the turn of the century, König, the brilliant student of Helmholtz, explained what was wrong in the eyes of dichromats, i.e. those extreme types of red/green colour defective who can match red and green lights exactly by adjusting intensity only. These defectives are of two kinds: deuteranopes to whom a red light is bright, and protanopes who see it dim. König explained this by supposing that deuteranopes lack the green-sensitive cone pigment, protanopes lack the red-sensitive. With only one cone pigment in the red/green spectral range, only one colour can be appreciated (Principle of Univariance), and lack of the red-sensitive pigment in the protanope makes red lights look dim to him. An alternative good explanation is Fick's, that dichromats possess the *normal* cone pigments, but that the red and green signals are not properly processed to give red/green discrimination. Retinal densitometry has decided in favour of König.

If the light reflected from the normal eye is measured using lights of various wavelengths, (a) in the dark-adapted state and (b) after bleaching with strong red light, a difference spectrum is obtained. Since the red bleaching light removes chiefly the red-absorbing pigment, it will cause an increase in light reflected through the retina chiefly in the red spectral range, i.e. the difference spectrum will be greatest in the red. Similarly after a blue-green bleach the difference spectrum will be greatest in the green. This expectation is in fact obtained in the normal eye. And it would be obtained also in the eyes of dichromats if Fick were right and they had both normal pigments. If on the contrary König were right and they had only one cone pigment active in the red/green spectral range, then only one kind of difference spectrum could be obtained, and red and green bleaching lights would give identical results if they were adjusted in intensity so that each bleached 50 % of the total pigment. This latter result is in fact obtained; thus we confirm König. Protanopes lack red-sensitive cones, deuteranopes lack green-sensitive cones.

The importance of confirming König's conjecture is that it leads to an accurate measure of the spectral sensitivity of red-sensitive and green-sensitive cones. Protanopes and deuteranopes can make accurate matches with light of various wavelengths against a fixed yellow. This gives us the spectral sensitivity of the single, but different, cone pigment that each possesses.

SINGLE CONES

The difference spectrum of the pigment in single cones from an excised retina has been measured by Marks, Dobelle and MacNichol, by

Liebmann and by Haroshi. This is a *tour de force* since the measuring light fast bleaches away the minute amount of pigment to be measured. However, with human cones this brilliant work confirms measurement in living dichromats and shows that no cones contain a pigment mixture.

LAND'S 2-COLOUR PROJECTIONS

Land has produced some very striking demonstrations as follows. First he obtained two black–white photographic positives after photographing a coloured scene (a) through a filter that transmitted only wavelengths longer than sodium light and (b) through a filter that transmitted only shorter wavelengths. These two plates, identical as to form but differing as to black/white density, were placed as slides in two projectors, and the pictures were superimposed exactly in register on a screen.

In one beam was interposed a pink transparency, in the other a grey neutral filter to make the uncoloured picture about as bright as the pink. Clearly every region of the screen is being lit by a pink light diluted with white in a range of brightness and dilutions.

The appearance is unbelievable!

The beautiful red-head in the green dress on the screen in Land's first demonstration at the American National Academy of Sciences took my breath away. I rolled up my programme and looked through the tube at isolated areas. Land looked at us and said 'All right, look through your tubes, it won't make any difference to the colours.' Nor did it.

On another occasion Land answered two criticisms (a) that the eye, by looking at this mainly pink picture, got colour-adapted and lost true colour judgement; and (b) this was abetted by including in his pictures oranges, etc. that would carry a clue to colour interpretation.

Land showed a picture of a green mound on which a white flagstaff displayed the stars and stripes. This was presented by an enormously powerful electronic flash that lasted less than 1 msec. 'Ready!' Bang! And we saw the picture in its colours. 'What did you see? You didn't have long to get *adapted* did you? Look again.' The projectionist had now interchanged the pink and neutral transparencies so that the picture would appear in approximately complementary colours. Bang! 'What did you see?' Awed voices murmured, 'The stripes were green!' 'Do you say that because you know so well what colour to expect?'

It certainly is not easy to know what is going on in the eye. A few years ago an engineer in Cambridge got very interested in Land's projections and invited several of us over from Physiology to see them.

With the usual pink and white projection lights the picture exhibited

in one place a very convincing green. 'Yes,' said Horace Barlow, 'very
remarkable, very strange how this presentation affects motor speech.'
'Motor speech! What on earth do you mean?' 'Well, what you have on
the screen there is a mixture of pink and white and what you have just
said is not pale "pink" but "green". Obviously something has gone
wrong with your speech.' 'Nothing is wrong with my speech, it actually *is*
green.' 'Quite so, the speech malfunction is consistent, you say it again.'
By now the engineer was trying not to burst, and further motor speech
was inhibited.

But Barlow's ingenious interpretation was wrong! I repeated Land's
projections with a special viewing arrangement in which I looked at the
'green' patch of the picture with my left eye, and simultaneously at the
corresponding place on the right retina I viewed a quite separate little
white card that could receive an adjustable mixture of monochromatic
lights – red plus green plus blue. By hand I varied the mixing knobs until
the mixture seen by the right eye was identical in colour with Land's
'green' seen by the left eye. The mixture was found to contain predominantly honest green light of wavelength 540 nm.

This could not be due to malfunction of motor speech, for nothing had
been said.

What about colour contrast, colour illusions and the surprising appearances of colour with various kinds of intermittent stimulation? Colour
opponent processes, first stressed by Hering and analysed in a long series
of psycho-physical researches by Hurvich and Jameson, have received
support from electrophysiology of the retina and lateral geniculate body
(DeValois and Gouras). The colour-coded horizontal cells of Svaetichin
and MacNichol are organized on a basis of colour contrast (Naka) and
mediate the 'cross-talk' between the output of cones of different colour
sensitives (Fuortes, Baylor, Simon).

What out of all this is colour, that *femme fatale* we see with our eyes and
in our minds and get so emotional about? Newton in his first paper said,
'But to determine more absolutely what light is and by what modes or
actions it produceth in our minds the phantasms of colours is not so easie.'
It certainly is not just the ratio of quantum catches in Young's three types
of cone.

Every night I cycle home from Newton's College under the sodium lights
which shine above indisputably yellow. Below is the painted white traffic
line in the middle of the road looking indisputably white. These two luminous areas, surrounded by black, both emit nothing but sodium radiation;
it is yellow above, and white below. *What indeed is colour?*

THE GASTROINTESTINAL HORMONES:
AN HISTORICAL REVIEW

By R. A. GREGORY

THE DISCOVERY OF SECRETIN

It has been said that the great discoveries of science are those which can be seen – sometimes long afterwards – to have changed our way of thought about natural phenomena and so to have turned the course of discovery in a new and fruitful direction which none had earlier foreseen. By this criterion, the discovery in 1902 by Bayliss and Starling, at University College London, of the duodenal hormone secretin was indeed a signal event in the history of physiology. A simple experiment, the work of a single afternoon, revealed that the functions of the body were normally co-ordinated not only by the nervous system, but also by the mediation of specific chemical agents formed in, and transmitted from, one organ to others by way of the circulation, conveying a message intelligible only to those cells equipped to capture the 'chemical messenger' and decipher the encoded instructions for modification of their activity. By the discovery of what came to be called 'hormones' there was opened a new era of physiology, the beginning of endocrinology as we know it today.

To understand how the discovery of secretin came about, it is necessary to look at the events which led up to it, which took place not in London but elsewhere in Europe and above all in St Petersburg. There, during the last twelve years or so of the nineteenth century, much of the foundations of modern gastroenterology was laid by the studies of Pavlov and his pupils on the work of the digestive glands. The functions of the digestive system had preoccupied Pavlov from his early days. When in 1929 he visited Montreal and was shown round the physiological laboratory of McGill University, he took from a shelf in the library volume I of George Henry Lewes's *The Physiology of the Common Life* (1859) opened it at page 230 and showed to his companions a diagram of the alimentary tract. 'When in my very young days I read this book in a Russian translation,' he said, 'I was greatly intrigued by this picture. I asked myself: how does such a complicated system work? My interest in the digestive system originated at that time' (Babkin, 1949). Early in his career Pavlov came under the influence of S. P. Botkin, according to whose doctrine of 'nervism' most of the bodily functions were regulated by the nervous system; and in the working of the digestive glands Pavlov saw a new and fruitful field for the study of nervous influences. He realized at the outset that his experiments must be made as far as possible in the conscious animal; and being endowed with great surgical skill he perfected methods for the surgical preparation of his dogs so that the secretions of the various digestive glands could be studied daily in the same conscious healthy animal.

Pavlov was first concerned to establish beyond doubt the secretory innervation to the pancreas and stomach, about which there was still uncertainty. In 1888, using a conscious dog provided with a pancreatic fistula and one vagus nerve divided in the neck some days previously (so as to give time for the cardiac fibres to degenerate) he stimulated the peripheral end of the nerve and obtained from the fistula a free flow of pancreatic juice. The next year, with his life-long assistant Madame Schumov-Simanovskaia, he performed his famous experiment of 'sham-feeding' a conscious dog provided with an oesophageal fistula and a gastric fistula. This resulted in a prompt secretion of gastric juice rich in acid, mucus and pepsin, which was prevented by previous vagotomy or the administration of atropine. These two experiments appeared at the time to have established beyond further question the role of the vagus in causing secretion from the pancreas and stomach and the nature of its effect on the secretory cells of those organs. However, in 1894 Pavlov's pupil Dolinski found that the introduction of dilute hydrochloric acid into the duodenum of a conscious dog provided with a pancreatic fistula caused a profuse and watery secretion of pancreatic juice. As a matter of fact, this discovery had already been made as early as 1825 by Leuret and Lassaigne (Mutt, 1959) but their work had been overlooked. Dolinski was perhaps led to his experiment by the observation of Bekker (1893) that carbonated water introduced into a dog's duodenum stimulated pancreatic secretion. It was at once apparent that in Pavlov's previous experiment on the pancreas, stimulation of the cervical vagus would also have excited gastric secretion, and the acid passing into the duodenum must have been at least partly responsible for the observed flow of pancreatic juice. In 1896 Pavlov again stimulated the cervical vagus, this time in an acute experiment on a dog in which the pyloric canal was occluded by a plug of cotton wool soaked in bicarbonate solution so as to prevent entry of acid into the duodenum but not to damage the vagal fibres which ran across the pyloric sphincter to the duodenum and thence to the pancreas. The result was very different from that previously observed – a slow flow of viscid juice rich in enzyme, which evidently represented the true effect in the dog of direct vagal excitation to the pancreas.

Thus there arose the problem of the nervous pathways followed by 'Dolinski's reflex', as it was called, the totally unexpected solution to which was to come from the hands of Bayliss and Starling six years later. Pavlov nearly stumbled on the answer; in 1897 he considered the possibility that acid absorbed into the circulation from the duodenum might be the active agency, but rejected this because acid introduced into the rectum (from which it was presumed to have been absorbed) did not excite pancreatic secretion. Attempts by others to define the nervous pathways concerned in the supposedly reflex effect proved fruitless; for instance Popielski (1901), a former pupil of Pavlov's, showed that acid in the duodenum still stimulated pancreatic secretion after section of the vagi and splanchnics, destruction of the medulla and spinal cord, removal of the coeliac ganglia and transection of the pylorus; he concluded that short reflex pathways must exist between the duodenum and the ganglion cells present in the pancreas itself. Wertheimer & Le Page (1901) made similar denervation experiments with the same result. They also showed that the intravenous injection of hydrochloric acid did not stimulate pancreatic secretion; and observing that atropine, though it blocked the action of the vagus and of pilocarpine, did not affect the response to acid in the duodenum, concluded that the efferent pathways of the reflex must be sympathetic in character, rather than vagal. They then went to the brink of the great discovery by showing that pancreatic secretion was excited on introduction of acid *into a loop of jejunum resected from the rest of the intestine*; they concluded from this that the centre for the reflex was situated in the ganglia of the solar plexus.

At this point in the story there are no better words than those of Bayliss & Starling (1902b)

but they [*Wertheimer & Le Page*] did not perform the obvious control experiment of injecting acid into an isolated loop of jejunum after extirpation of these ganglia. The apparently local character of this reaction interested us to make further experiments on the subject in the idea that we might have here to do with an extension of the local reflexes whose action upon the movements of the intestines we have already investigated. We soon found, however, that we were dealing with an entirely different order of phenomena and that the secretion of the pancreas is normally called into play not by nervous channels at all but by a chemical substance which is formed in the mucous membrane of the upper parts of the small intestine under the influence of acid and is carried by the blood stream into the gland cells of the pancreas.

The 'local reflexes' already studied by Bayliss and Starling referred to their investigation on the mechanism of the intestinal movements (Bayliss & Starling, 1899) in which they had shown that local stimulation of the intestine, as by a distending bolus, excited a dual response of reflex nature, the pathways of which lay in the myenteric plexuses, resulting in contraction behind and relaxation before the point of stimulation. This dual response of opposite sign transmitted down the gut in the plexuses they named 'peristalsis'.

The work they were now to recount in detail had been briefly described in a preliminary communication (Bayliss & Starling, 1902a) to the Royal Society on 23 January 1902; it was entitled 'On the causation of the so-called peripheral reflex secretion of the pancreas'. The first experiment was made on 16 January, and their friend Sir Charles Martin later described what had taken place (Martin, 1927).

I happened to be present at their discovery. In an anaesthetised dog a loop of jejunum was tied at both ends and the nerves supplying it dissected out and divided so that it was connected with the rest of the body only by its blood vessels. On the introduction of some weak HCl into the duodenum, secretion from the pancreas occurred and continued for some minutes. After this had subsided a few cubic centimetres of acid were introduced into the denervated loop of duodenum. To our surprise a similarly marked secretion was produced. I remember Starling saying 'Then it must be a chemical reflex'. Rapidly cutting off a further piece of jejunum he rubbed its mucous membrane with sand and weak HCl, filtered and injected it into the jugular vein of the animal. After a few moments the pancreas responded by a much greater secretion than had occurred before. It was a great afternoon.

Bayliss & Starling (1902b) continued,

Since Wertheimer and Le Page had shown that the effect of acid in the small intestine diminishes in proportion as the place where it is introduced approaches the lower end, so that from the last six inches or so of the ileum no secretion of the pancreas is excited, it was of interest to see whether the distribution of the substance . . . is similar in extent.

An extract made from the lower ileum in the same way as the jejunal extract had no effect on the pancreas; but since both extracts caused a similar fall in blood pressure it was thus established that the effect on pancreatic secretion was not due merely to vasodilatation, but to an agent located only in that region of the intestinal mucosa from which the pancreatic response to acid could be obtained.

In our supposedly fast-moving times it is salutary to note that by the time Bayliss and Starling came to write the full account of their work only a few months after the preliminary note to the Royal Society, they could refer to several publications by others which had already appeared on the subject. Among them was a demonstration by Wertheimer that the blood coming from a loop of intestine into which essence of mustard had been introduced was capable of exciting pancreatic secretion when infused intravenously into another dog; and also an objection by Pflüger to their interpretation of their findings. He argued that the denervation of the intestinal loop in the original experiment could not have been complete because of nerves running within the walls of the mesenteric vessels. Their reply was,

We admit that it is difficult to be certain that all nerve channels were absolutely excluded . . . but we submit that since the result of the experiment was such as has been demonstrated it does not in the least matter whether the nerves were all cut or not; the only fact of importance is that it was the *belief* that all the nerves were cut that caused us to try the experiment of making an acid extract from the mucous membrane, and that led to the discovery of secretin. Exit Pflüger!

Bayliss and Starling were unable to repeat successfully Pavlov's demonstration that vagal stimulation would excite pancreatic secretion, and stated, 'In our opinion the chemical mode of excitation, viz. by the production of secretin in the mucous membrane by the action of the acid chyme from the stomach upon it, is the normal one. At all events, this mode of stimulation must take place whether there is a concomitant nervous process or not, so that this latter is superfluous and therefore improbable.' Their view was eventually changed by the visit to University College London in 1912 of G. V. Anrep, a pupil of Pavlov's, who demonstrated the experiment to them; they had failed because their dogs were always given morphine as a preoperative sedative, and the drug causes spasm of the pancreatic duct system. On the other hand Bayliss and Starling's observations were easily confirmed in St Petersburg, where they had a significant outcome. Pavlov asked his assistant Savich to perform the experiment, and Babkin (1949) who was present later recounted the scene. 'The effect of secretin was self-evident. Pavlov and the rest of us watched the experiment in silence. Then without a word Pavlov disappeared into his study. He returned half an hour later and said, "Of course, they are right. It is clear that we did not take out an exclusive patent for the discovery of the truth" '. The doctrine of 'nervism' in the affairs of the digestive system was obviously dead, and Pavlov later remarked to Babkin, 'Of course we may continue to study with success the physiology of digestion, but let other people do it. As for myself, I am getting more and more interested in the conditioned reflexes.' In 1904 when Pavlov received the Nobel Prize for his studies on the work of the digestive glands his acceptance speech was largely concerned with the nature of appetite, the 'psychic' stimulation of secretion, and conditioned reflexes. In 1913 and 1914 the discovery of secretin was declared by the Nobel examiner to be worthy of a prize, but the war intervened; and when Starling was again nominated in 1926 (Bayliss had died in 1924), the work was thought too old to be eligible (Liljestrand, 1952).

The wider significance of the discovery of the 'messenger' role of secretin was discussed by Bayliss & Starling (1904) in their joint Croonian Lecture to the Royal Society and by Starling (1905) in his Croonian Lectures to the Royal College of Physicians, entitled 'The chemical correlation of the functions of the body'. In the first lecture he remarked,

These chemical messengers, however, or 'hormones' from the Greek ὁρμάω (as we might call them) have to be carried from the organ where they are produced to the organ which they affect by means of the blood stream, and the continually recurring physiological needs of the organism must determine their repeated production and circulation throughout the body.

This was the first use of the word 'hormone' and of its choice Bayliss (1915) later said,

When we came across the mode by which the pancreas was excited to activity it became obvious to Starling and myself that the chemical agent concerned was a member of a class of substances of which others were previously known. The group . . . is characterised by the property of serving as *chemical messengers*. They enable a chemical correlation of the functions of the organism to be brought about through the blood side by side with that which is the function of the nervous system. This being so, it seemed desirable and convenient to possess a name to distinguish the group. That of 'internal secretion' already in use did not sufficiently emphasise their nature as messengers. Finally Mr W. B. Hardy proposed the name of 'hormone' derived from ὁρμάω (I arouse to activity) and although the property of messenger was not suggested by it, it was adopted. It has in fact been generally understood as having the meaning intended and not to be applied to any kind of substance which excites activity.

Needham (1936) says that according to local tradition the word was born in the hall of Caius College, Cambridge. 'Schäfer or Starling was brought in to dine by Hardy and the question of nomenclature was raised. W. T. Vesey, an authority on Pindar, suggested ὁρμάω and the thing was done.'

GASTRIN

The first major consequence of the discovery of secretin was the discovery of gastrin. In the third of his Croonian Lectures (27 June 1905) Starling said,

In the alimentary canal itself the chemical correlation between intestine and pancreas does not stand alone . . . Edkins has shown that a secondary secretion of gastric juice is determined by the production of a hormone in the pyloric part of the mucous membrane (of the stomach) under the influence of the first products of digestion, and that this hormone is absorbed by the blood and carried by it to the gastric glands in the fundus, which are thereby excited to renewed activity.

Five weeks previously J. S. Edkins had read to the Royal Society a preliminary communication entitled 'On the chemical mechanism of gastric secretion' (Edkins, 1905). It began:

It has long been known that the introduction of certain substances into the stomach provokes a secretion of gastric juice . . . On the analogy of what has been thought to be the mechanism at work in the secretion of pancreatic juice by Bayliss and Starling, it is probable that in the process of absorption of digesting food in the stomach a substance may be separated from the cells of the mucous membrane which, passing into the blood or lymph, stimulates the secretory cells of the stomach to functional activity.

He went on to describe simple experiments on anaesthetized cats in which the intravenous injection of aqueous extracts of pyloric mucosa stimulated gastric secretion, while similar extracts made from the fundic mucosa did not; the heat-stable active principle he named 'gastrin'.

Edkins was no novice in the study of digestion, and long before the discovery of secretin he was thinking about the problem of how gastric secretion was continued in the later stages of gastric digestion, after the initial vagal reflex demonstrated by Pavlov had presumably come to an end. To volume I of Schäfer's *Textbook of Physiology* (1898) he contributed a chapter entitled 'Mechanism of secretion of gastric, pancreatic and intestinal juices' in which he cited the finding of Heidenhain (1879) that food placed in the main stomach of a conscious dog caused secretion in a vagally denervated pouch of the gastric fundus (the earliest type of gastric pouch, which Heidenhain had invented the year before). This fact is now recognized to constitute powerful evidence that gastric secretion is hormonally stimulated; Heidenhain had concluded that certain products of digestion absorbed from the stomach excited the secretion. Edkins remarked, 'If it is absorbed digestion products that provoke secretion, is it a specific product or products that cause this to occur, or is it a common characteristic of all?' Quoting some experiments of Chishin in Pavlov's laboratory showing that peptone was particularly effective in exciting gastrin section, Edkins went on: 'We may assume that small quantities of peptone may be normally formed in the stomach, and becoming absorbed there in some way influence the epithelium (of the gastric glands) so that secretion results.' Elsewhere in the chapter he had discussed the vague ideas of the day concerning the role of the pyloric glands, and the supposed presence in them of pepsin, concluding, 'It yet remains to be discovered whether the cells of the pyloric glands possess other more important functions.' With the discovery of secretin his ideas crystallized. The function of the pyloric region was to absorb the products of gastric digestion, notably peptone; and there was carried with them into the blood stream a hormone stored in the pyloric glands. This is no doubt why in his later search for gastrin he made his extracts of pyloric mucosa with solutions of peptone, dextrose and maltose.

Edkins's triumph was short-lived. Within a few years it was being shown by others that aqueous extracts made from a variety of organs would stimulate gastric secretion when injected subcutaneously or intramuscularly into conscious dogs provided with gastric fistulae or pouches. The discovery of histamine in intestinal mucosa (Barger & Dale, 1911), the recognition of its presence in every organ in the body, and the demonstration by Popielski (1919) that it was a most potent stimulant of gastric acid secretion, led to the view that Edkins's pyloric extract owed its power to stimulate gastric secretion to the presence of this substance, which because of its ubiquitous distribution could hardly be included in the select group of 'chemical messengers' as originally defined by Bayliss and Starling. Only Lim (1922) repeated Edkins's experiments exactly as he had performed them and concluded that he had been right; there was a stimulant distinct from histamine in the pyloric mucosa. Lim's work went unheeded, and the explanation of the puzzle was only to come after the isolation of gastrin (Gregory, 1970); meanwhile the negative results of apparently well-conceived physiological tests of the 'gastrin theory' and a failure to find in pyloric mucosal extracts any stimulant of gastric secretion other than histamine fostered the general belief that a gastric hormone did not exist. Even the discovery by Komarov (1938) in Montreal that a gastric secretagogue of protein character separable from histamine was present in pyloric extracts did little to influence contemporary opinion, for 'Komarov's gastrin' as it was called appeared to be effective only in the anaesthetized cat, where its action was totally resistant to atropine, which was well known to inhibit the response to a meal in conscious animals and man. In the end this resistance to atropine in the anaesthetized, though not the conscious, animal proved to be one of the remarkable properties of gastrin; Komarov's extract, like Edkins's, had indeed contained the hormone. After more than forty years of general disbelief, the existence of an antral hormone was proven by physiological experiment (Grossman, Robertson & Ivy, 1948); it was isolated in 1962 (see Gregory, 1962), identified as a heptadecapeptide amide in 1963 and its total synthesis accomplished in 1964 (Gregory, 1970).

Meanwhile, over the years, secretin had become the subject of a vast number of physiological and chemical studies; but the hormone itself had defied all attempts a capture. This was achieved almost exactly sixty years after its discovery by the Swedish chemists Erik Jorpes and Viktor Mutt, who isolated it from porcine duodenum and identified it as a strongly basic peptide of twenty-seven amino acid residues (Jorpes, Mutt, Magnusson & Steele, 1962). Elucidation of the sequence, completely different from that of gastrin, and total synthesis of the hormone, followed in 1966 (Jorpes, 1968). What has been called the 'biochemical era' in the study of the gastrointestinal hormones had begun.

THE OTHER HORMONES

The discovery of secretin and what is now recognized to have been the discovery of gastrin inevitably prompted attempts to establish the existence of other gastrointestinal hormones. In most cases these efforts followed the pattern laid down by Bayliss and Starling's classical experiment; a physiological response associated with a meal, hitherto regarded as a nervous reflex, was examined for the presence of a hormonal component by looking for the persistence of the effect after the interruption of nervous connexions between the site of origin and the target organ. Attempts were then made to prepare a mucosal extract which would reproduce the physiological effect upon injection and so might be considered to contain the putative hormonal principle. It is of interest to review these studies in order to trace the subsequent history of the ideas involved.

INCRETIN

In point of time it might be said that the idea of this hormonal action originated perhaps even earlier than the idea of a gastric hormone. In December 1902 Starling was reviewing, in a paper given to the Pathological Society of London, the pathological implications of recent work on the pancreas (see Starling, 1903), and speaking of secretin and its actions, he said,

Since diabetes appears to be connected in some way with the pancreas, we thought it possible that some effect might be produced on the disease by intravenous injections of solutions of secretin. This hope however proved to be unfounded, Dr. Spriggs (of the London Hospital) having tried the intravenous injection of secretin in a case of diabetes without producing any effect whatever on the course of the disease.

There is no further reference to this idea in the writings of Bayliss or Starling; but at that time, or soon afterwards, Dale began to study in Starling's laboratory the effect of a secretin extract on the islets of Langerhans, which were suspected to be the source of the internal secretion of the pancreas. Dale's histological observations seemed to support the view already expressed by others that the islets were a phase in the life-cycle of the acinar cells, for the number of islets appeared to increase on prolonged stimulation with secretin (Dale, 1904). These ideas did not pass unnoticed; Moore, Edie & Abram (1906) tried the effect of a secretin extract (orally) in cases of diabetes, on 'the hypothesis that the *internal* secretion of the pancreas might be stimulated . . . by a substance of the nature of a hormone or secretin yielded by the duodenal mucous membrane . . .'. From Dale's results, they surmised

that the pancreas contains but one type of secreting cell which yields both the internal and external secretion, and that the cells of the islets of Langerhans are ordinary pancreatic cells in a phase of exhaustion. If this be the case, the likelihood is increased that anything which stimulates the external secretion will also stimulate the internal secretion.

Others failed to confirm their belief that secretin improved the diabetic state (Bainbridge & Beddard, 1906).

In the 1920s, after the discovery of insulin, there appeared many papers suggesting that a hypoglycaemic agent was present in duodenal mucosa. A leading figure in this field was La Barre (1936), who suggested that the secretin molecule was a complex consisting of 'excretin', which stimulated pancreatic exocrine secretion, and 'incretin', which stimulated the internal secretion, i.e. insulin. The claims of a duodenal hypoglycaemic hormone failed to survive critical examination (Best, Jephcott & Scott, 1932; Loew, Gray & Ivy, 1940); but the careful studies of Laughton & Macallum (1932) can now be seen to have provided substantial support for what would be regarded today as an 'incretin' effect. They prepared a secretin-free duodenal extract which did not lower the fasting blood sugar in a normal dog, or in a totally pancreatectomized dog, so that it did not contain an agent which was itself hypoglycaemic, i.e. insulin-like. However, it did markedly diminish the hyperglycaemia produced in normal animals by the intravenous injection of glucose or the hyperglycaemia produced by feeding glucose to partially pancreatectomized animals. They concluded,

As to the mode of action of our preparation it would appear that the most probable explanation lies in the assumption that the preparation stimulates the islets of Langerhans to secrete insulin. If this is correct there must be a very delicate balance or the insulin if in excess would tend to produce a hypoglycaemia. Our observations gave no evidence that this occurs.

After a long interval, interest in the existence of a duodenal factor in insulin release was revived by suggestions that the oral administration of glucose accelerated the rate of disappearance of a subsequent intravenous load, and that for a given rise in blood glucose, blood insulin-like activity was higher when the glucose was given intraduodenally than when it was injected intravenously (Arnould, Bellens, Franckson & Conard, 1963).

Finally McIntyre, Holdsworth & Turner (1964, 1965), who measured plasma insulin levels by the specific and sensitive method of radioimmunoassay, demonstrated unequivocally in human subjects that for the same glucose load the blood sugar curve was lower, and the plasma insulin curve higher, when it was given intrajejunally than when it was given intravenously (Fig. 1); and in human subjects also, Dupré (1964) showed that a crude commercial secretin extract accelerated the disappearance of an intravenous glucose load. There was thus established the existence of a duodenal agent of hormonal character which controlled insulin release. The problem that remained was the identity of 'incretin'.

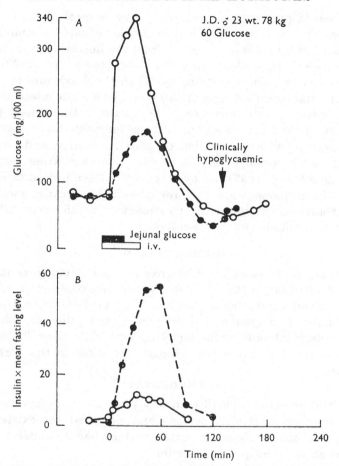

Fig. 1. The effects on (*A*) blood glucose and (*B*) plasma insulin levels of intrajejunal (●) and intravenous (○) glucose in a human subject. For the same glucose load (60 g) the blood glucose curve is lower and the plasma insulin curve higher for intrajejunal administration than for intravenous infusion. (From McIntyre *et al.*, 1965.)

ENTEROGASTRONE

Ewald & Boas (1886), introducing to clinical gastroenterology the idea of a test-meal for the study of gastric secretion using the newly invented stomach tube, discovered that olive oil added to a meal of starch paste given to human subjects inhibited gastric secretion and delayed gastric emptying. Pavlov and his pupils studied the effect (considered to be a reflex) and showed that it resulted from the presence of fat not in the stomach itself, but in the duodenum. Farrell & Ivy (1926) discovered by accident that in a conscious dog provided with a completely transplanted gastric pouch (a preparation invented by Ivy) a meal containing fat inhibited tone and motility; they had in fact anticipated a stimulation of motility claimed by Le Heux to result from the liberation of an intestinal 'motor hormone'. Their observation led Feng, Hou & Lim (1929) to show inhibition of gastric acid secretion, by fat in the duodenum in a similar preparation; and Kosaka & Lim (1930) prepared a crude intestinal extract which inhibited gastric secretion. They named the active principle 'enterogastrone' and there were speculations later that there might be two enterogastrones, one for inhibition of gastric secretion and one for inhibition of gastric motility. Efforts to purify enterogastrone, notably by Gray, Bradley & Ivy (1937) and by others after them proved fruitless and interest in the hormone gradually waned despite the great attraction of its possible therapeutic value in the treatment of peptic ulcer, which had been the main stimulus to their studies.

CHOLECYSTOKININ

Ivy & Oldberg (1928) established by cross-circulation experiments in dogs that the effect of fat in the duodenum in causing contraction of the gall bladder involved a humoral mechanism. They made active extracts from porcine duodenal mucosa and distinguished the principle, which they named 'cholecystokinin', from secretin; but while their findings were generally accepted little further interest was taken in the hormone for many years.

ENTEROCRININ

Nasset (1938) detected a small secretory response in denervated jejunal loops in conscious dogs after feeding and proposed the existence of a hormonal mechanism. He made extracts which had a similar action and named the active principle 'enterocrinin'.

VILLIKININ

The observation that motor activity of the intestinal villi is stimulated by a meal led Kokas & Ludany (1934) to demonstrate that hydrochloric acid introduced into the duodenum of a dog increased villus activity in a jejunal loop temporarily transplanted into the neck of the animal. They made extracts of intestinal mucosa which stimulated villus motility and were considered to contain a hormone 'villikinin'.

PANCREOZYMIN

The discovery of pancreozymin may be fairly regarded as the third major advance made in knowledge of the control of pancreatic secretion after the discovery of the effect of the vagus by Pavlov and of secretin by Bayliss and Starling. From the time of Pavlov there had accumulated a number of observations on the stimulation of pancreatic enzyme secretion by food-stuffs, particularly protein and fat placed in the duodenum, which were not satisfactorily explained in terms of the reflex activity supposed to be involved; and during the many unsuccessful attempts to purify secretin there arose a controversy as to its action on pancreatic enzyme secretion. The preparation made by Mellanby did not, and he attributed enzyme secretion entirely to the action of the vagus. On the other hand the secretin made by American workers according to a different procedure undoubtedly did stimulate enzyme production. There the matter rested until Harper & Vass (1941), who were examining in anaesthetized cats the effect on pancreatic secretion of vagal and splanchnic stimulation and of introducing protein into the small intestine using a background flow of enzyme-poor juice produced by Mellanby's secretin extract, discovered that the stimulation of enzyme output produced by casein in the intestine persisted after section of the extrinsic nerves. They recognized the implications of this observation; and Harper & Raper (1943) found a potent stimulant of pancreatic enzyme secretion, which they named 'pancreozymin', in a side-fraction discarded during the preparation of secretin by Mellanby's method. They separated the principle from secretin and demonstrated that its action, unlike that of the vagus, was resistant to atropine in the anaesthetized animal. Later the elegant experiments of Wang & Grossman (1951) provided physiological evidence for the existence of a duodenal hormone stimulating pancreatic enzyme secretion using conscious dogs provided with a portion of the pancreas completely transplanted to the mammary region.

The chemical identity of pancreozymin remained unknown for more than twenty years after its discovery until Jorpes and Mutt in 1964, having completed their work on secretin, turned their attention to the cholecystokinin and pancreozymin activities present in their crude duodenal extract. During purification the two activities remained inseparable and there was eventually isolated a single basic peptide of thirty-three amino acids which proved to be a powerful stimulant both of gall bladder contraction and of pancreatic enzyme secretion (Jorpes, 1968). The same hormone had been discovered twice on the basis of what are now recognized to be two of its principal actions, and it has come to be generally agreed that it should be known as 'cholecystokinin' (CCK) after the action by which its existence was first recognized.

Gastric inhibitory polypeptide (GIP)

Brown & Pederson (1970) noted that two partially purified preparations of CCK provided by Jorpes and Mutt had inhibitory effects on gastric acid secretion which were quantitatively different from their potency for gall bladder contraction. The inhibitor was isolated by Brown, Mutt & Pederson (1970) and proved to be a peptide of forty-three amino acids (Brown & Dryburgh, 1971). GIP is a potent inhibitor of gastric acid secretion and motility; radioimmunoassay shows that it is released from the duodenum by feeding, particularly of glucose and fat. GIP has also been found to possess 'incretin' activity (Brown, Dryburgh, Ross & Dupré, 1975) and to stimulate intestinal secretion (Barbezat, 1973).

THE PHYSIOLOGICAL ACTIONS OF THE GASTROINTESTINAL HORMONES

In all of the searches made for new gastrointestinal hormones after the discoveries of secretin and gastrin, there was a general assumption that the hormone sought had a single physiological action, that by which its existence had been first recognized. Although Bayliss and Starling (Starling, 1906) had noted that their secretin preparation also stimulated the flow of hepatic bile and had attributed this to the action of secretin (Fig. 2), and Edkins regarded gastrin as a stimulant of gastric pepsin as well as of acid, there was never seriously considered thereafter the likelihood that one hormone might have more than one physiological action. Neither was there considered the less obvious possibility that more than one hormone might prove to have the same physiological action. It came as something of a surprise when pure gastrin was shown to have several actions besides that of stimulating gastric acid secretion (Gregory & Tracy, 1964); it also stimulated the secretion of gastric pepsin and pancreatic enzyme, it contracted the gall bladder (weakly) and it stimulated gastrointestinal tone and motility. Subsequent studies with the pure natural or synthetic hormone extended this list to include several other actions, including the growth of gastric mucosa and the stimulation of insulin release.

When pure secretin became available the story was repeated; it not only stimulated the secretion of water and bicarbonate from the pancreas and from the liver (as Bayliss and Starling had supposed) but it had a number of other actions also; for instance it inhibited gastric secretion and motility, it stimulated pepsin secretion, the release of insulin, the secretion of Brunner's glands and lipolysis of fat cells, and it antagonized the trophic actions of gastrin. The isolation of CCK and elucidation of its structure established it as a member of the 'gastrin family', for the C-terminal tetrapeptide amide was identical with that of gastrin, already shown to be the minimal totipotent active fragment of the molecule; and as was expected the range of actions of CCK proved to be closely similar to that of gastrin, although certain differences in the structure of the C-terminal heptapeptide of CCK compared with gastrin conferred upon it

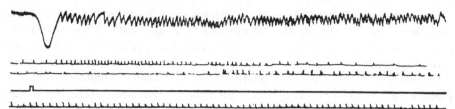

Fig. 2. Effect of intravenous injection of a bile-salt-free secretin extract on the flow of pancreatic juice and hepatic bile in an anaesthetized dog. Tracings from above downwards: (1) blood pressure; (2) drops of pancreatic juice; (3) drops of bile; (4) signal marking injection of secretin extract; (5) 10-sec intervals. (From Starling, 1906.)

the characteristic activity of a most powerful action upon the gall bladder. The resemblance of GIP to secretin identified it as a member of the 'secretin family'; it shared with secretin the power to inhibit gastric acid secretion and motility and to release insulin, and in addition it stimulated intestinal secretion.

As these multiple actions of pure gastrin, secretin, CCK and GIP were revealed in turn, it became increasingly clear that there were candidates among them for the physiological roles attributed to 'incretin', 'enterogastrone', and 'enterocrinin'. They could all be shown to cause insulin release in experiments *in vitro* or *in vivo*; secretin and GIP inhibited gastric secretion and motility, while GIP stimulated intestinal secretion. However, all these actions had been discovered and studied in the first place by the use of the pure peptides in doses which had no known relation to 'physiological limits'. It was obvious from the outset that some actions could be elicited at very low dosage rates, whereas others seemed to require much greater amounts for their demonstration; but what was quite uncertain was how the blood levels produced in such experiments compared with those which occurred in natural circumstances following a meal; which of the many actions found for each hormone could be regarded as of physiological significance? The advent of radioimmunoassay offered the prospect of a sensitive and specific method for determining the circulating levels of the hormones, and so promised an approach to this fundamental problem by (1) measuring the postprandial circulating level of a hormone and then (2) reproducing it in the same animal on another occasion by infusion of the pure hormone, with observation of the actions which might follow on the various target organs. As will be seen, this apparently simple procedure turns out to be far from straightforward in practice. Some information of this nature has appeared in respect of gastrin and GIP but the radioimmunoassays for secretin and CCK are not yet sufficiently refined to enable them to be used for this purpose.

For instance infusions of gastrin which produce increases in plasma levels considered to be within the physiological range have been shown to stimulate acid secretion and also to increase the plasma insulin response provoked by an intravenous glucose load – the 'incretin' effect (Rehfeld & Stadil, 1973). GIP inhibits gastrin-stimulated acid secretion when infused so as to produce circulatory increases which are within the range observed after feeding a meal. Infusion of GIP plus glucose (in man) increases the rise in plasma insulin observed after glucose alone; the plasma level of GIP is within the range observed after oral administration of glucose (Dupré, Ross, Watson & Brown, 1973).

Assuming for the moment that the radioimmunoassay used gives a true picture of the plasma hormone content, studies of this simple pattern clearly provide vital information; but there will remain uncertainty about those actions of a given hormone which may not be demonstrable by this means. The reasons for failure may include the following: (1) The response of the digestive tract to a meal involves a component of vagal activity as well as the effects of the various hormones and the action of a hormone may be dependent upon concurrent vagal activity. An obvious example of such dependency is that of gastrin on the oxyntic cell, which is greatly potentiated by concurrent vagal excitation. On the other hand it appears from present evidence that vagal excitation to the pancreas is not an important factor in the response of that organ to secretin and CCK. (2) A co-operative interaction at a given target organ may exist between the various hormones liberated by a meal so that giving only one of them in the amount observed postprandially has little or no effect on the target organ. A notable example of this is the interaction of secretin and CCK on pancreatic secretion. Acid in the duodenum is the only known effective releaser of secretin, but acidification of the duodenum to the same degree and extent as is observed following a normal meal produces a volume-response from the pancreas that is far less than that observed in the same animal after a normal meal. Clearly there is a missing factor in the response and this appears to be CCK, which is normally released at the same time as secretin by components such as fat and protein rather than by acid. It can be shown by infusion of the two hormones separately or together into a dog provided with a pancreatic fistula that the presence of CCK greatly potentiates the volume-response produced by secretin. Conversely, it can be similarly shown that secretin increases the stimulation of pancreatic enzyme secretion by CCK. It may thus be necessary to infuse two or more hormones so as to produce for each plasma levels which are within postprandial limits. There is, however, a more formidable problem to be faced that derives from the heterogeneity of the hormone measured by the radioimmunoassay. The nature of the problem and how it may be dealt with are well exemplified by gastrin (Gregory, 1974).

The hormone was first isolated from antral mucosa in the form of a peptide having seventeen amino acids (G17) which appeared to account for virtually all of the gastrin-like activity present in the tissue. The C-terminal tetrapeptide was the active region of the molecule; antibodies raised against conjugated G17 reacted with the C-terminal region and were made the basis of a radioimmunoassay for the hormone. However, in 1970 the use of such a radioimmunoassay in conjunction with the fractionation of plasma by molecular sieving on Sephadex columns showed that although material corresponding to G17 was present, the predominant amount of immunoreactivity corresponded to a larger form of the hormone ('Big' gastrin). On brief digestion with trypsin BG disappeared and its place was taken by an equivalent amount of G17, from which it was surmised that BG might consist of G17 covalently linked through an arginine or lysine residue (the points of attack of trypsin) to a further peptide chain. In 1972 there were isolated from porcine antral mucosa and from human gastrinomas peptides having thirty-four amino acids (G34) which corresponded in immunological and chromatographic behaviour to plasma BG; the C-terminal portion of the molecule was G17, joined by two lysine residues to a further peptide chain.

In antral tissue G34 accounts for no more than about 5 % of the total gastrin present, almost all of the remainder being G17. G34 predominates in plasma chiefly because it has a half-life several times longer than G17, but its potency for stimulation of gastric acid secretion (based on acid responses to equal plasma levels) is several times less than that of G17, and it is the latter which makes the major contribution to postprandial gastric acid responses. Clearly an accurate description of the relationship between postprandial acid secretion and plasma gastrin level measured by radioimmunoassay requires the measurement of both forms of the hormone simultaneously, and this cannot easily be done by the use of a single antibody, which is likely to react with both forms of the hormone. It can, however, be accomplished by the use of two antibodies, which show different reactivities to the two forms (Dockray & Taylor, 1976). The possible presence in the circulation of active forms of the hormone too small to react significantly with the antibodies used for radioimmunoassay cannot be excluded; but there is as yet no strong evidence that such exist.

So far as the presence in circulation of biologically inactive but immunologically reactive forms or fragments of the hormone is concerned there has already been discovered by the use of a double antibody assay, material which corresponds to the inactive N-terminal tridecapeptide of G17, and it is possible that the inactive fragment which distinguishes G34 from G17 may also circulate; but it is not now difficult to take account of this because the availability of many fragments and forms of gastrin make it possible to characterize with a high degree of precision the reactivity of an antibody before it is used for radioimmunoassay.

For various reasons, the assays for the other hormones have not yet attained the status of that for gastrin in sensitivity, reliability and validation; and it is therefore not yet clear what may be the nature and degree of the heterogeneity of them in circulation, although there have been reports of more than one circulating form of all of them. In the tissue of origin (intestinal mucosa) there is only one report so far of heterogeneity; Jorpes and Mutt have isolated CCK-Variant, six residues longer at the N-terminus than CCK. From what has been said it seems clear that there is a long road to be travelled before the circulating active forms and amounts of the other hormones besides gastrin can be described precisely in terms of postprandial plasma levels and this knowledge applied to define the full range of their truly physiological actions.

THE CANDIDATE HORMONES

Few would take serious issue today with the view that secretin, gastrin, cholecystokinin and gastric inhibitory peptide can be regarded on the basis of the available evidence as 'established' hormones. For each there is at least one physiological action by which their existence was first recognized and which has been shown to be humorally transmitted; they have each been isolated and chemically characterized, and the pure peptide has been shown to be capable of reproducing the physiological effect attributed to the hormone. Their location in the gastrointestinal mucosa is confined to those regions from which the effects attributed to them can be obtained by application of an appropriate stimulus associated with a meal. Thus, gastrin is confined to the pyloric and upper intestinal mucosa, from which regions the hormonal components of the 'gastric' and 'intestinal' phases of a meal response take their origin; the others are located in the duodenal and jejunal regions and similar physiological evidence associates their distribution with their hormonal role. Radioimmunoassay shows that after feeding, immunoreactivity corresponding to gastrin, GIP and CCK appears in the peripheral circulation; this demonstration has not yet been achieved for secretin, most probably owing to inadequate sensitivity of the assay in its present form. However, there is powerful indirect evidence that the hormone does appear in the peripheral circulation following an ordinary meal (Fig. 3) and it has been demonstrated even with present assays that immunoreactivity appears there following acidification of the

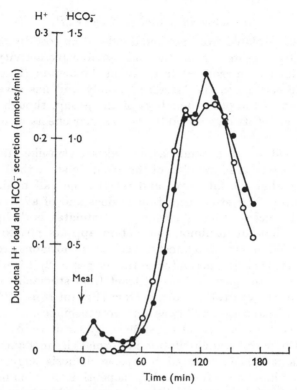

Fig. 3. Conscious pancreatic fistula dog; a test-meal is emptying into the duodenum. Pancreatic secretion (bicarbonate output (●)) runs closely parallel to the quantity of H^+ (○) entering from the stomach. (From Moore, Verine & Grossman, 1976.)

duodenum, albeit in somewhat greater extent than normally occurs after feeding.

The great and increasing interest in this field of gastrointestinal endocrinology during the past few years has brought forth a growing number of what have been aptly termed 'candidate hormones' (Grossman *et al.*, 1974; Grossman, 1975); they range from physiologically active pure peptides (some of them already well known from elsewhere in the body) whose possible hormonal role in the affairs of the gastrointestinal tract is uncertain or altogether problematical, to physiological actions of demonstrably hormonal character for which no corresponding mucosal principle has yet been identified. Some members of this group deserve brief discussion here, if only to indicate the many growing points in this area.

Vasoactive intestinal peptide (*VIP*)

This peptide was isolated from porcine duodenal mucosa by Said & Mutt (1972) on the basis of its vasodilator and hypotensive activity; a similar peptide has since been isolated from avian duodenum (Nilsson, 1974). Structurally it belongs to the 'secretin' family and has several actions similar to those of the other members of the group, though it also has distinctive effects of its own, notably the vascular effects which led to its identification.

It inhibits acid secretion, and has a weak secretin-like effect on pancreatic secretion (the only member of the group to show this). Like secretin, it increases hepatic bile flow, and relaxes the gall bladder. It is a potent stimulant of intestinal secretion in dogs and of adenylate cyclase in rabbit ileal mucosa. Like glucagon, it stimulates both lipolysis and glycogenolysis; but an insulinotropic action appears not to have been demonstrated. What argues against a peripheral hormonal role for VIP are the facts that (1) it is inactivated in the liver and (2) it is widely distributed throughout the gastrointestinal tract from stomach to colon, and is also found in the central nervous system (Bryant *et al.*, 1976); it may thus have some local role, possibly as a neurotransmitter. A radioimmunoassay for VIP is in use, but is not yet sufficiently sensitive to show whether it appears in the peripheral circulation in normal circumstances, although elevated levels have been reported in hepatic cirrhosis, suggesting that it may be released into the portal blood. This peptide has been implicated in the causation of some clinical conditions of watery diarrhoea, notably the Vernier–Morrison syndrome, in which severe watery diarrhoea, hypokalaemia and achlorhydria is associated with a pancreatic tumour which apparently secretes large amounts of the peptide.

Enteroglucagon

Glucagon-like biological activity was discovered in extracts of dog gastrointestinal mucosa by Sutherland & De Duve (1948) and glucagon-like immunological activity (GLI) was identified there by Unger *et al.* (1961). It was virtually confined to the mucosal layer and its distribution was wide, with high concentrations in the gastric fundus, jejunum and ileum. GLI can be differentiated from pancreatic glucagon by site-specific antibodies raised against the latter, and this has been made the basis of a radioimmunoassay for it. Using such an assay, attempts have been made at isolation but without complete success so far; there is evidence that part of the material is identical with pancreatic glucagon and is distributed differently from the remainder, which is chromatographically highly heterogeneous. By radioimmunoassay, GLI has been shown to be released into the circulation after feeding, particularly of carbohydrate and fat, and its most firmly established activity at present appears to be glycogenolysis.

Pancreatic polypeptide (PP)

This candidate for hormonal status is of particular interest because its discovery exemplifies a new turn in the searches for further gastro-intestinal hormones. With so many gastrointestinal hormones now known to influence so many of the gastrointestinal functions associated with diges-tion, it obviously becomes increasingly difficult to identify a new hormone on the basis of its physiological effect; but instances are now appearing in which, as Grossman (1975) has put it, the traditional course of events has been stood on its head; a peptide is isolated first and its possible hormonal role examined by radioimmunoassay afterwards.

During the purification of glucagon and insulin, there were isolated from pancreatic extracts of cow, hog, sheep, man and chicken, new homo-logous peptides (PP) containing thirty-six amino acids (Lin & Chance, 1974; Kimmel, Hayden & Pollock, 1975). These have been shown to have a wide variety of effects on gastrointestinal secretory and motor functions in experimental animals. Cells apparently containing the peptide have been demonstrated by immunofluorescence not only in the pancreatic islets but also in small groups between the acinar cells; and following a meal in man there is a prompt release of PP immunoreactivity into the circulation lasting several hours (Adrian *et al.*, 1976). This release appears to be dependent upon vagal excitation, since it is greatly decreased in vagotomized patients (Schwartz *et al.*, 1976). Reproducing the post-prandial plasma levels by infusion of the pure peptide should throw some light on the problem of its possible hormonal status; the functional significance of PP is at present totally unknown.

THE HORMONE RECEPTORS

At the heart of that great afternoon's work in 1902, when Bayliss and Starling began it all with their discovery of secretin, was their recogni-tion of the 'messenger' role of hormones, with its implication of the 'recognition' of the circulating hormone molecule by a specific structure, the 'receptor' possessed by the 'target' cell. It has become a funda-mental aim of endocrinology to understand at the molecular level the nature of the interaction which takes place between the hormone and its receptor; and for its full achievement this requires ultimately the isolation and physicochemical characterization of the receptor 'molecule', using that term to describe the complex involved in the translation of 'recog-nition' into cellular response. The primary event in recognition is the reversible binding of the hormone molecule to the receptor site, which in the case of peptide hormones is generally agreed to be located on the cell surface. In recent years, this process has been increasingly studied in several favourable situations, where two essential prerequisites can be satisfied. These are (1) radiolabelling of the hormone without loss of physiological activity, and (2) the preparation of viable populations of target cells, or of their plasma membranes (Cuatrecasas, 1974). For instance, the reactions of glucagon and insulin with their receptors on liver or fat-cell membranes have been studied by Rodbell, Birnbaumer, Pohl & Sundby (1971), insulin and growth hormone receptors on cultured human lymphocytes have been examined by Gavin, Gorden, Roth, Archer & Buell (1973) and Lesniak, Roth, Gorden & Gavin (1973), and the interactions of glucagon, enteroglucagon, VIP, and secretin at their receptors on liver and fat-cell membranes have been described by Bataille, Freychet & Rosselin (1974).

In a few instances, highly purified preparations have been made of hormone-binding macromolecules which are believed to represent the receptor. Thus, Cuatrecasas (1972) purified by procedures involving affinity chromatography the insulin receptor of liver-cell membranes to a point which was considered to approach theoretical purity; this involved a concentration of nearly 500 000-fold, so minute was the amount of receptor in the starting material.

Such fundamental approaches to the problem of the complex hormone–receptor interactions at the major gastrointestinal target cells, e.g. the oxyntic and pancreatic acinar cells, would be of the greatest value; and this area of study will no doubt develop rapidly as the present problems associated with hormone radiolabelling and preparation of viable homogeneous cell populations are surmounted. Two studies of great future promise have recently been reported: (1) Amsterdam & Jamieson (1972) succeeded in isolating guinea-pig pancreatic acinar cells in a state of excellent viability; they were capable of incorporating radiolabelled amino acids into enzyme protein and of releasing this in response to secretagogues such as carbaminoycholine, CCK or caerulein added to the incubation medium. This preparation was used by Klaeveman, Conlon & Gardner (1975) to obtain plasma membranes, which they used to study the interactions at their receptor sites of CCK-octapeptide, VIP, secretin, gastrin and glucagon, as indicated by the changes in activity of membrane-bound adenylate cyclase, which mediates the action of many hormones. As the authors recognized, it would have been ideal to have examined the interactions of these hormones directly, by studying the binding of them in a radiolabelled form, but they were unable at that time to achieve labelling without inactivation. Nevertheless, it was shown for instance that the receptor for CCK-octapeptide, with which gastrin also interacted, was

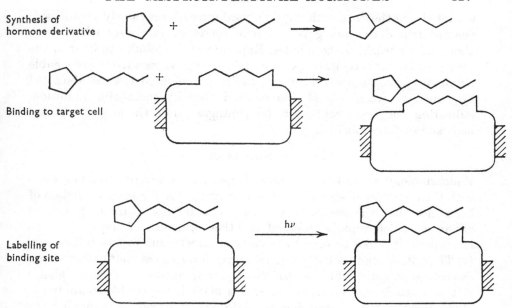

Fig. 4. Photoaffinity labelling of a peptide hormone-binding site: (1) a peptide hormone derivative is prepared which upon photoactivation will covalently bond to a protein. (2) The hormone derivative interacts with its receptor site on the cell plasma membrane. (3) Photolysis results in the formation of a covalent bond between the hormone derivative and protein in the receptor site. (From Galardy & Jamieson, 1975.)

functionally distinct from the receptor with which VIP and secretin interacted. (2) A novel and potentially fruitful approach to the problem of characterizing morphologically and functionally, and perhaps of isolating, the receptor sites for gastrointestinal hormones, is that of photoaffinity labelling.

A peptide hormone derivative is prepared which retains its physiological activity, but which upon photoactivation will covalently bond to protein. The hormone derivative is then allowed to interact with, and so bind to, its receptor site by addition in the dark to a preparation of viable target cells or their plasma membranes. Finally, exposure of the system to light results in photolysis of the hormone derivative with the formation of a covalent bond between the hormone derivative and a receptor site protein (Fig. 4). Galardy & Jamieson (1975) prepared a photoactivatable derivative of pentagastrin (an active analogue of the C-terminal pentapeptide amide of gastrin), namely 2-nitro-5-azidobenzoyl pentagastrin, and added this to a preparation of surviving pig pancreatic lobules (small clusters of acini) which was capable of discharging radiolabelled secretory protein in response to stimulation by CCK or pentagastrin. Incubation of the lobules in the dark with unphotolysed (or with previously photolysed) pentagastrin derivative gave maximal secretory responses which were abolished by washing the lobules. Exposure of the lobules to light in the presence of photoactivatable pentagastrin derivative resulted in irreversible maximal stimulation of secretion; the response could not be abolished by extensive washing and was blocked only by metabolic inhibition, indicating that the receptor site for pentagastrin on the acinar cells had been successfully labelled.

CONCLUSION

Endocrinology as we know it today began nearly seventy-five years ago with the discovery of secretin and recognition of the messenger function of hormones; but the gastrointestinal system did not share in the great advances which rapidly followed in other branches of the subject. In retrospect, this can be seen to have been largely due to two difficulties: (1) The cells of origin of the gastrointestinal hormones are not gathered into discrete aggregations but are widely dispersed among the exocrine glands of the stomach and small intestine; this made it impossible to apply the classical approach of studying an endocrine function by removing the gland of origin, and it also made more difficult the attempts to identify an endocrine principle by extraction of it from its site of origin, because of the great amounts of extraneous material present. (2) The hormones themselves proved to be peptides of small or moderate size, present in low concentration, and of such nature as to be easily lost or inactivated by the extraction procedures of the time. These problems eventually found their solutions; and since 1962 the isolation and identification of the major gastrointestinal hormones, together with the application of radioimmunoassay which this made possible, has resulted in a remarkable advance of knowledge in every aspect of this field of study. It has been rightly said (A. G. E. Pearse) that the gastrointestinal tract is proving to be the largest and most complex endocrine gland in the body; and there would seem to be a certain justice in this outcome to the many barren years, since, after all, it was there that it all began.

REFERENCES

ADRIAN, T. E., BLOOM, S. R., BRYANT, M. G., POLAK, J. M. & HEITZ, P. H. (1976). Radioimmunoassay of a new gut hormone – human pancreatic polypeptide. *Gut* **17**, 393–394.
AMSTERDAM, A. & JAMIESON, J. D. (1972). Structural and functional characterisation of isolated pancreatic exocrine cells. *Proc. natn. Acad. Sci. U.S.A.* **69**, 3028–3032.
ARNOULD, Y., BELLENS, R., FRANCKSON, J. R. M. & CONARD (1963). Insulin response

and glucose-C^{14} disappearance rate during the glucose tolerance test in the un-anaesthetised dog. *Metabolism* 12, 1122–1131.

BABKIN, B. P. (1949). *Pavlov. A Biography.* Chicago: University of Chicago Press.

BAINBRIDGE, F. A. & BEDDARD, A. P. (1906). Secretin in relation to diabetes melli-tus. *Biochem. J.* 1, 429–441.

BARBEZAT, G. O. (1973). Stimulation of intestinal secretion by polypeptide hor-mones. *Scand. J. Gastroenterol.* 8, *Supplement* 22, 1–21.

BARGER, G. & DALE, H. H. (1911). β-Iminazolylethylamine, a depressor constituent of intestinal mucosa. *J. Physiol.* 41, 499–503.

BATAILLE, D., FREYCHET, P. & ROSSELIN, G. (1974). Interactions of glucagon, gut glucagon, vasoactive intestinal polypeptide and secretin with liver and fat cell membranes; binding to specific sites and stimulation of adenylate cyclase. *Endo-crinology* 95, 713–720.

BAYLISS, W. M. (1915). *Principles of General Physiology.* London: Longmans, Green.

BAYLISS, W. M. & STARLING, E. H. (1899). The movements and innervation of the small intestine. *J. Physiol.* 24, 99–143.

BAYLISS, W. M. & STARLING, E. H. (1902a). On the causation of the so-called 'Peripheral Reflex Secretion' of the pancreas. *Proc. R. Soc.* 69, 352–353.

BAYLISS, W. M. & STARLING, E. H. (1902b). The mechanism of pancreatic secretion. *J. Physiol.* 28, 325–353.

BAYLISS, W. M. & STARLING, E. H. (1904). Croonian Lecture: The chemical regula-tion of the secretory process. *Proc. R. Soc.* 73, 310–322.

BEKKER, N. M. (1893). Zur pharmakologie der Alkalien. Diss., St Petersburg. Quoted by Babkin, B. P. (1928). *Die Äussere Sekretion Der Verdauungsdrüsen,* 2nd edn, p. 502. Berlin: Springer.

BEST, C. H., JEPHCOTT, C. M. & SCOTT, D. A. (1932). Insulin in tissues other than the pancreas. *Am. J. Physiol.* 100, 285–294.

BROWN, J. C. & DRYBURGH, J. R. (1971). A gastric inhibitory polypeptide. II. The complete aminoacid sequence. *Can. J. Biochem.* 49, 867–872.

BROWN, J. C., DRYBURGH, J. R., ROSS, S. A. & DUPRÉ, J. (1975). Identification and actions of gastric inhibitory polypeptide. *Rec. Prog. Hormone Res.* 31, 487–532.

BROWN, J. C., MUTT, V. & PEDERSON, R. A. (1970). Further purification of a poly-peptide demonstrating enterogastrone activity. *J. Physiol. Lond.* 209, 57–64.

BROWN, J. C. & PEDERSON, R. A. (1970). A multiparameter study on the action of preparations containing cholecystokinin–pancreozymin. *Scand. J. Gastroenterol.* 5, 537–541.

BRYANT, M. G., BLOOM, S. R., ALBUQUERQUE, R. H., POLAK, J. M., MODLIN, I. & PEARSE, A. G. E. (1976). Possible dual role for vasoactive intestinal peptide as gastrointestinal peripheral hormone and neurotransmitter substance. *Lancet* i, 991–993.

CUATRECASAS, P. (1972). Affinity chromatography and purification of the insulin receptor of liver cell membranes. *Proc. natn. Acad. Sci. U.S.A.* 69, 1277–1281.

CUATRECASAS, P. (1974). Membrane receptors. *A. Rev. Biochem.* 43, 169–214.

DALE, H. H. (1904). On the 'Islets of Langerhans' in the pancreas. *Phil. Trans. R. Soc.* B 197, 25–46.

DOCKRAY, G. J. & TAYLOR, I. L. (1976). Heptadecapeptide gastrin: measurement in blood by specific radioimmunoassay. *Gastroenterology* (in press).

DUPRÉ, J. (1964). An intestinal hormone affecting glucose disposal in man. *Lancet* ii, 672–673.

DUPRÉ, J., ROSS, S. A., WATSON, D. & BROWN, J. C. (1973). Stimulation of insulin secretion by gastrin inhibitory polypeptide in man. *J. clin. Endocrinol. Metab.* 37, 826–828.

EDKINS, J. S. (1905). On the chemical mechanism of gastric secretion. *Proc. R. Soc.* B **76**, 376.

EWALD, C. A. & BOAS, J. (1886). Beiträge zur Physiologie und Pathologie der Verdauung. *Virchows Arch. path. Anat. Physiol.* **104**, 271–305.

FARRELL, J. I. & IVY, A. C. (1926). Studies on the motility of the transplanted gastric pouch. *Am. J. Physiol.* **76**, 227.

FENG, T. P., HOU, H. C. & LIM, R. K. S. (1929). On the mechanism of the inhibition of gastric secretion by fat. *Chinese J. Physiol.* **3**, 371–380.

GALARDY, R. E. & JAMIESON, J. D. (1975). Photoaffinity labelling of secretagogue receptors in the pancreatic exocrine cell. In *Gastrointestinal Hormones* (Symposium), ed. THOMPSON, J. C., pp. 345–366. Austin and London: University of Texas Press.

GAVIN, J. R., GORDEN, P., ROTH, J., ARCHER, J. A. & BUELL, D. N. (1973). Characteristics of the human lymphocyte insulin receptor. *J. biol. Chem.* **248**, 2202–2207.

GRAY, J. S., BRADLEY, W. H. & IVY, A. C. (1937). On the preparation and biological assay of enterogastrone. *Am. J. Physiol.* **118**, 463–476.

GREGORY, R. A. (1962). Gastric secretion: A review of its chief nervous and hormonal mechanisms. In *Surgical Physiology of the Gastrointestinal Tract* (Symposium), ed. SMITH, A. N., pp. 57–70. Edinburgh: Royal College of Surgeons.

GREGORY, R. A. (1970). Gastrin – the natural history of a peptide hormone. *Harvey Lectures Series* **64**, 121–155.

GREGORY, R. A. (1974). The Bayliss–Starling Lecture 1973. The gastrointestinal hormones: a review of recent advances. *J. Physiol.* **241**, 1–32.

GREGORY, R. A. & TRACY, H. J. (1964). The constitution and properties of two gastrins extracted from hog antral mucosa. Part I. The isolation of two gastrins from hog antral mucosa. Part II. The properties of two gastrins isolated from hog antral mucosa. *Gut* **5**, 103–117.

GROSSMAN, M. I. (1975). Additional candidate hormones of the gut (Letter). *Gastroenterology* **69**, 570–571.

GROSSMAN, M. I., ROBERTSON, C. R. & IVY, A. C. (1948). The proof of a hormonal mechanism for gastric secretion – the humoral transmission of the distension stimulus. *Am. J. Physiol.* **153**, 1–9.

GROSSMAN, M. I. et al. (1974). Candidate hormones of the gut. *Gastroenterology* **67**, 730–755.

HARPER, A. A. & RAPER, H. S. (1943). Pancreozymin, a stimulant of the secretion of pancreatic enzymes in extracts of the small intestine. *J. Physiol.* **102**, 115–125.

HARPER, A. A. & VASS, J. N. (1941). The control of the external secretion of the pancreas in cats. *J. Physiol.* **99**, 415–435.

HEIDENHAIN, R. (1979). Über die Absonderung der Fundusdrüsen des Magens. *Pflügers Arch. ges. Physiol.* **19**, 148–169.

IVY, A. C. & OLDBERG, E. (1928). A hormone mechanism for gallbladder contraction and evacuation. *Am. J. Physiol.* **86**, 599–613.

JORPES, J. E. (1968). Memorial Lecture. The isolation and chemistry of secretin and cholecystokinin. *Gastroenterology* **55**, 157–164.

JORPES, J. E., MUTT, V., MAGNUSSON, S. & STEELE, B. B. (1962). Aminoacid composition and N-terminal aminoacid sequence of porcine secretin. *Biochem. bipohys. Res. Commun.* **9**, 275–279.

KIMMEL, J. R., HAYDEN, L. J. & POLLOCK, H. G. (1975). Isolation and characterisation of a new pancreatic polypeptide hormone. *J. biol. Chem.* **251**, 9369–9376.

KLAEVEMAN, H. L., CONLON, T. P. & GARDNER, J. D. (1975). Effects of gastrointestinal hormones on adenylate cyclase activity in pancreatic exocrine cells.

In *Gastrointestinal Hormones* (Symposium), ed. THOMPSON, J. C., pp. 321–344. Austin and London: University of Texas Press.

KOKAS, E. & LUDANY, G. (1934). Die hormonale Regelung der Darmzottenbewegung II. *Pflügers Arch. ges. Physiol.* **234**, 182–186.

KOMAROV, S. A. (1938). Gastrin. *Proc. Soc. exp. Biol. Med.* **38**, 514–516.

KOSAKA, T. & LIM, R. K. S. (1930). Demonstration of the humoral agent in fat inhibition of gastric secretion. *Proc. Soc. exp. Biol. Med.* **27**, 890–891.

LA BARRE, J. (1936). *La Sécrétine. Son Rôle Physiologique, ses Propriétés Thérapeutiques.* Paris: Bibliothèque Scientifique Belge, Section Biologique & Masson et Cie.

LAUGHTON, N. B. & MACALLUM, A. B. (1932). The relation of the duodenal mucosa to the internal secretion of the pancreas. *Proc. R. Soc.* B **111**, 37–46.

LESNIAK, M. A., ROTH, J., GORDEN, P. & GAVIN, J. R. (1973). Human growth hormone radioreceptor assay using cultured human lymphocytes. *Nature New Biol.* **241**, 20–21.

LEWES, G. H. (1859). *The Physiology of the Common Life.* Edinburgh & London: Blackwood & Sons.

LILJESTRAND, G. (1952). The Nobel Prize in Physiology and Medicine. In *Nobel, The Man and his Prizes*, pp. 135–316. Stockholm: the Nobel Foundation, Sohlmans Förlag.

LIM, R. K. S. (1922). The question of a gastric hormone. *Quart. Jl exp. Physiol.* **13**, 79–103.

LIN, T. M. & CHANCE, R. E. (1974). Gastrointestinal actions of a new bovine pancreatic peptide. In *Endocrinology of the Gut*, ed. CHEY, L. Y. & BROOKS, F. P. Thorofare, New Jersey: Charles B. Slack Inc.

LOEW, E. R., GRAY, J. S. & IVY, A. C. (1940). Is a duodenal hormone involved in carbohydrate metabolism? *Am. J. Physiol.* **129**, 659–663.

MCINTYRE, N., Holdsworth, C. D. & TURNER, D. S. (1964). New interpretation of oral glucose tolerance. *Lancet* ii, 20–21.

MCINTYRE, N., HOLDSWORTH, C. D. & TURNER, D. S. (1965). Intestinal factors in the control of insulin secretion. *J. clin. Endocrinol.* **25**, 1317–1324.

MARTIN, C. J. (1927). Obituary Ernest Henry Starling. *Br. med. J.* **1**, 900–905.

MOORE, B., EDIE, E. S. & ABRAM, J. H. (1906). On the treatment of diabetes mellitus by acid extract of duodenal mucous membrane. *Biochem. J.* **1**, 28–38.

MOORE, E. W., VERINE, H. J. & GROSSMAN, M. I. (1976). The duodenum is an integrator and an amplifier: H^+ ion load drives pancreatic bicarbonate secretion. *Am. J. Physiol.* (in press).

MUTT, V. (1959). On the preparation of secretin. *Arkiv för Kemi* **15**, 75–95.

NASSET, E. S. (1938). Enterocrinin, a hormone which excites the glands of the small intestine. *Am. J. Physiol.* **121**, 481–487.

NEEDHAM, J. (1936). *The Terry Lectures. Order and Life.* London: Cambridge University Press.

NILSSON, A. (1974). Isolation, amino acid composition and terminal amino acid residues of the vasoactive octacosapeptide from chicken intestine. Partial purification of chicken secretin. *FEBS Lett.* **47**, 284–289.

POPIELSKI, L. (1901). Über das peripherische reflektorische Nervenzentrum des Pankreas. *Pflügers Arch. ges. Physiol.* **86**, 215–224.

POPIELSKI, L. (1919). β-imidazolyläthylamin und die Organextrakte. I. β-imidazolyläthylamin als mächtiger erreger der Magendrüsen. *Pflügers Arch. ges. Physiol.* **178**, 214–259.

REHFELD, J. F. & STADIL, F. (1973). The effect of gastrin on basal- and glucose-stimulated insulin secretion in man. *J. clin. Invest.* **52**, 1415–1426.

R. A. GREGORY

RODBELL, M., BIRNBAUMER, L., POHL, S. L. & SUNDBY, F. (1971). The reaction of glucagon with its receptor: evidence for discrete regions of activity and binding in the glucagon molecule. *Proc. natn. Acad. Sci. U.S.A.* **68**, 909–913.

SAID, S. I. & MUTT, V. (1972). Isolation from porcine-intestinal wall of a vasoactive octacosapeptide related to secretin and to glucagon. *Eur. J. Biochem.* **28**, 199–204.

SCHÄFER, E. A. (1898). *Textbook of Physiology*, vol. I. Edinburgh and London: Young J. Pentland.

SCHWARTZ, T. W., REHFELD, J. F., STADIL, F., LARSSON, L.-I., CHANCE, R. E. & MOON, N. (1976). Human pancreatic polypeptide response to food in duodenal ulcer patients before and after vagotomy. *Lancet* **i**, 1102–1105.

STARLING, E. H. (1903). On some pathological aspects of recent work on the pancreas. *Trans. pathol. Soc. Lond.* **54**, 253–258.

STARLING, E. H. (1905). The chemical correlation of the functions of the body. *Lancet* **ii**, 339–341.

STARLING, E. H. (1906). *Mercer's Company Lectures on Recent Advances in the Physiology of Digestion.* London: Constable.

SUTHERLAND, E. W. & DE DUVE, V. (1948). Origin and distribution of hyperglycemic–glycogenolytic factor of pancreas. *J. biol. Chem.* **175**, 663–674.

UNGER, R. H., EISENTRAUT, A. M., SINKS, K., McCALL, M. S. & MADISON, L. L. (1961). Sites of origin of glucagon in dogs and humans (abstr). *Clin. Res.* **9**, 53.

WANG, C. C. & GROSSMAN, M. I. (1951). Physiological determination of release of secretin and pancreozymin from intestine of dogs with transplanted pancreas. *Am. J. Physiol.* **164**, 527–545.

WERTHEIMER, E. & LE PAGE, L. (1901). Sur l'association reflexe du pancreas avec l'intestin grêle. *J. Physiol. Path. Gén.* **2**, 689–692.

PERINATAL PHYSIOLOGY

BY R. A. McCANCE

1600–1900

The study of adults has usually preceded that of infants. Great discoveries, however, in perinatal anatomy and physiology had been made by William Harvey in his *De generatione* and by De Graff (1668, 1672), Boyle (1670) and Mayow in the seventeenth century (see Parsons, 1950). Galen had preceded these men by many years and Jenner, John Hunter, Lavoisier and Cavendish followed them in the eighteenth century. By the middle of the nineteenth century percussion, and auscultation, introduced by Auenbrugger and Laennec, respectively, had enormously advanced the study of normal and abnormal cardiac and pulmonary function, and made it possible to diagnose the presence of ascites during life. These discoveries were to some extent lost on the professional physiologists of the day.

Work was going forward on chemical embryology, particularly that of the avian egg (Needham, 1931). Nevertheless, little progress had been made in the study of prenatal and neonatal physiology although, as in adult physiology, studies of infantile form and structure had already begun, and comparisons with adults could be made. A number of these were assembled and summarized by Vierort (1888; and see Bernard, 1865). In the first half of the nineteenth century the proteins, fats and carbohydrates were being separated and knowledge about their role in nutrition and metabolism was making great strides (Moleschott, 1859).

By the 1850s, therefore, the stage was set for Claude Bernard to make his great contribution to perinatal physiology. 'Le choix intelligent d'un animal présentant une disposition . . . heureuse est souvent la condition essentielle du succès d'une expérience et de la solution d'un problème physiologique très important' (Bernard, 1865). The discovery had been made ten years before (Bernard, 1855) when he found that the fluids of the foetal calf contained quite large amounts of fructose (Bernard, 1855, p. 379 and see Bacon & Bell, 1948; Barklay *et al.*, 1949; Goodwin, 1956; Huggett, 1959). In the 1840s, with the assistance of two chemists Pelouse and Barreswil, Bernard had been working on glycogen formation and metabolism by the liver in foetal and adult life. He knew that if sucrose was injected intravenously or subcutaneously it was largely excreted unchanged in the urine. He identified the reducing substance which he found in the urine and foetal fluids as laevulose by (*a*) its behaviour with Barreswil's reagent and (*b*) its rotation of polarized light. He knew in addition that glucose could not be coming into the picture because by this point in gestation the liver had not yet reached the stage of development when it could make it (Bernard, 1855, pp. 253–254).

Bernard (1865) was well aware that a physician was always experimenting upon his patients in the hope that the conditions he was altering might cure or alleviate their sufferings. He recognized that the frog, even in those days, had probably contributed more to physiology than any other animal (*ibid.*, p. 201), but also that 'Il est bien certain que pour les questions d'application immédiate à la pratique médicale, les expériences faites sur l'homme sont toujours les plus concluantes' (*ibid.*, p. 215). Nevertheless Bernard was quite clear in his own mind what stand he should take up so far as human experimentation was concerned.

On a le devoir et par conséquent le droit de pratiquer sur l'homme une expérience toutes les fois qu'elle peut lui sauver la vie, le guérir où lui procurer un avantage personnel. La principe de moralité médicale…consiste donc à ne jamais pratiquer sur un homme une expérience qui ne pourrait que lui être nuisible à un degré quelconque bien que le résultat pût intéresser beaucoup la science, c'est-à-dire la santé des autres (*ibid.*, p. 176).

Some pioneering work was done on perinatal physiology in the second half of the nineteenth century. Pflüger (1868) and Zuntz (1877) showed that it was possible to keep a foetal sheep alive for some time, still attached to its mother by the umbilical cord, and to study the effect of clamping the cord at different stages of development. Renal function *in utero* was studied in a methodical but not very imaginative way by Dohrn (1867), who catheterized 100 newborn infants and made observations on the specific gravity and other characteristics of the urine, which have subsequently turned out to be correct. Similar observations were made on renal function before and after birth (Pollak, 1869) till the beginning of the twentieth century (Sabrazes & Fauquet, 1901); and later (Makepeace, Fremont-Smith, Dailey & Carroll, 1931; Wladimiroff & Campbell, 1974). Some of the observations made by Tausch (1936), however, are not in line with those of others.

The age of adventure

The closing years of the last century and the first 25 years of the present one were stamped by many experiments on man, usually made on the investigator himself but often on a colleague. High and low atmospheric pressures and dangerous mine gases were explored by Paul Bert (1878), repeatedly by J. S. Haldane (see Haldane & Priestley, 1935) and Joseph Barcroft (see Barcroft *et al.*, 1919–20; Barcroft, 1925, 1928, 1938) both in the laboratory and outside it (and see also Matthews *et al.*, 1954–55). J. B. S. Haldane and Priestley made daring attempts to alter the pH and volume of their *milieu intérieur* by drinking water or by taking toxic doses of ammonium chloride, calcium and strontium chlorides, sodium bicarbonate and also by overbreathing (Davies, Haldane & Kennaway, 1920; Haldane, 1921, 1925; Priestley, 1921; Haldane, Hill & Luck, 1923). These experiments made no direct impact on perinatal physiology at the time except where they interlocked with work being carried out earlier or at the same time on the passage of gases across the placenta.

It was these years too that saw so many discoveries beginning to be made about the so-called accessory food factors. Vitamins were being multiplied by the biochemists without any real appreciation of the effects they were to turn out to have on early growth and perinatal welfare. One might have expected, for instance, that the study of vitamins would have produced immediate information about the requirements for growth and thus led to its improvement more quickly than it did. After all, Hopkins had always emphasized the importance of a perfect diet for optimum growth, and it was a failure to grow on more and more highly purified diets that led to the search for and isolation of many of the vitamins, including vitamin C. People had not yet begun to think in terms of growth before and just after birth and what might be wrong with it. There were lots of reasons, after all, why infants should not grow satisfactorily, surrounded as they were in those days by a host of infections and dietary problems. No steps were taken about it, indeed, till Preston Maxwell (1930*a*, *b*) described foetal rickets in China, and later Liu *et al.* (1941) the extensive osteomalacia among pregnant women there. Bakwin (1937) had already noted the curious concentrations of calcium and phosphorus in serum in the first few days of life when infants were being reared on a cow's milk formula (Snelling, 1943; Gardner *et al.*, 1950; Gardner, 1952). By the onset of the second world war the impact of Mellanby's work had begun to sink in. 'Welfare' cod liver oil was being provided and National Dried Milk was fortified with vitamin D. Margarine had been fortified since 1940. Some time later, however, cases of hypercalcaemia began to come to light. A committee was appointed and reported (Committee, 1956) that the high serum calcium might be due to enthusiasm outrunning discretion, and stricter limits were set on the quantities of vitamin D to be added to infant foods. With the implementation of this, cases ceased to appear (Ćurčić & Ćurčić, 1974).

The fourth quarter of the last century saw the publication of some pioneer work on the passage of gases across the placenta (Pflüger, 1868, 1877; Zuntz, 1877; Cohnstein, 1884; Cohnstein & Zuntz, 1884). Zuntz used a variety of animals, including sheep.

Zuntz (1877) found that a rabbit foetus survived asphyxia longer than the mother and concluded therefore that per gram of body weight the foetus required less oxygen than the mother. Pflüger (1877) commented in the same year on the extraordinary tenacity of life exhibited by an early human embryo. Curiously enough it seems to have escaped both Pflüger and Zuntz that this applied to all mammalian foetuses and depended upon their ability to turn over to anaerobic metabolism, observations made long before by William Harvey and Robert Boyle.

The latter summarized his experiments thus: 'These tryals may deserve to be prosecuted with further ones, to be made not only with such kittens, but with other very young animals of different kinds; for by what has been related it appears, that those animals continued three times longer in the Exhausted Receiver, than other animals of that bigness would have done.'

Claude Bernard's (1865) views on what was an ethical experiment to carry out on a man have already been mentioned, and he enlarged upon these (ibid., p. 177). He knew quite well that experiments had at one time or another been carried out on criminals, and regarded an experiment made on a dying man as quite right and proper, provided it did not interfere with him in any way. He considered that Christianity and the medical ethics of the day allowed one to experiment on oneself or a colleague, but only if such experiments could not possibly do any harm. It was implied, therefore, that risks were not to be taken.

What then can one say of the experiments made by the Haldanes, Barcroft and others? We know at least what J. B. S. Haldane (1940) had to say about some of his own. The experiments were not entirely without risk and someone might have been held responsible had anything serious happened. The same may be said with even greater force about the experiments carried out between 1932 and 1939 by McCance and Widdowson. In them, accidents did indeed happen, caused by pyrogens, and after overbreathing, and a number of people were exposed to the chances of getting serum jaundice. None of these accidents would have happened with the skills available today. Yet every possible precaution, which the knowledge of the times suggested should be taken, was taken (McCance, 1935, 1936; McCance & Widdowson, 1936, 1939).

The first edition of Cushny's book (1917) had established by the early 1920s a unifying concept of renal function based upon the formation of a protein-free filtrate in the glomerulus, followed by considerable reabsorption from the tubules. According to this so-called 'Modern theory', substances not reabsorbed were classified as 'non-threshold' if they were not themselves actively secreted into the tubules. Chlorides and sodium were regarded as 'threshold' substances. It soon became clear that urea was not a 'non-threshold' substance and must leak back or be reabsorbed passively because creatinine was concentrated to a much greater extent. Rehburg (1926a, b) suggested, therefore, that for the time being at any rate creatinine should be regarded as one truly non-threshold substance and potentially a measure of the glomerular filtration rate, and McCance & Madders (1930) found that there was something to be said for using the pentose sugars for this purpose. This was demonstrated independently by Homer Smith's group (Clarke & Smith, 1932; Jolliffe, Shannon & Smith, 1932; Shannon, Jolliffe & Smith, 1932), but Höber (1933) and Shannon (1934, 1935) went on with this work and very shortly found that inulin was the best of the saccharide and polysaccharide complexes to use for measuring glomerular filtration rates, and it became for many years the standard material for doing so (Shannon & Smith, 1935).

THE PHYSIOLOGY OF DEVELOPMENT AND PERINATAL FUNCTION

The stage was now set for the discoveries that were to introduce a new era into the biochemistry of development and provide an almost explosive stimulus for neonatal physiology. The former was the confirmation by new semi-micro methods that there were foetal haemoglobins as distinct from adult haemoglobins in a number of species (Huggett, 1927; Brinkman & Jonxis, 1935, 1937; and see Brown, 1968; and Huehns & Beaven, 1971). The stimulus for neonatal physiology came from Winifred Young, who had gone to work with Sir Leonard Parsons in Birmingham after witnessing the whole of the work on salt deficiency which had been going on at King's College Hospital. Winifred Young noted with some concern when she was testing neonatal urines that they frequently contained no chlorides and wondered if the infants could possibly be salt deficient. An investigation was planned and carried out at Birmingham in eighteen normal newborn and older infants. Three infants with meningo-myeloceles were also investigated and on them glomerular filtration rates were measured with inulin. The urea clearances (see Austin, Stillman & Van Slyke, 1921; Möller, McIntosh & Van Slyke, 1929) were found to vary with the minute volume of the urine, but were always well below those of adults, whether the comparison was made on the basis of surface area, kidney weight or body weight. The sodium and chloride clearances were very low by adult standards even when the plasma values were abnormally high. The potassium clearances were low and the serum potassium tended

to be high, sometimes twice the adult level. The urines were always hypotonic and generally extremely so (McCance & Young, 1941). Premature
infants showed all these signs of immaturity more conspicuously than full-
term infants (Young, Hallum & McCance, 1941). By 1944 Heller had shown
that doses of posterior pituitary hormone which had a profound effect on
the diuresis of an adult scarcely altered the volume or concentration of the
urine of newborn infants. He further found (Heller, 1947; Heller &
Zaimis, 1949) that the posterior pituitary of the newborn rat and human
infant contained, respectively, only 1/10 and 1/5 as much hormonal
activity as that of the adult. At the back of one's mind during this work
there was always a worry about what one was justified in doing to normal
full-term and premature infants, even with the consent of their parents.
We felt happy enough about what we were doing when the procedures
were extensions of the established diagnostic and follow-up techniques of
the hospital. We did not, however, feel justified in determining a normal
infant's glomerular filtration rate, valuable though this would have been,
and it was for this reason that, after a consultation with Sir Leonard
Parsons, we decided to use infants with inoperable meningo-myeloceles if
the experiments were to involve catheterizing the babies and making
intravenous injections. There were no crippling ethical committees in
those days! The results with such children showed that the results of
the work on normal infants had been correct, and they were confirmed
in newborn infants (Barnett, 1940; Barnett, Perley & McGinnis, 1942)
and in premature infants (Barnett, Hare, McNamara & Hare, 1948a, b;
Barnett, McNamara, Hare & Hare, 1948). McCance & Wilkinson (1947)
showed that the administration by mouth of 5 % of the body weight of
water to adult and newborn rats was followed by a rapid and effective
diuresis in the adults, while the newborn animals produced no significant
diuresis and no dilution of the urine (compare also Dicker & Heller, 1945,
1951.) The administration of 2·5 % of the body weight as solutions of
10 % NaCl or 20 % urea led to an osmolar diuresis in the adult rat and a
fall in the osmolar concentration of the urine, whereas the same dose to a
newborn led only to a trifling diuresis and a small increase in the osmolar
concentration of the urine. Dean & McCance (1947a, 1949) found that the
response of newborn infants and adults to hypertonic solutions, given
intravenously, closely resembled those just described in rats, save that the
response of the newborn infants was relatively better than that of the
newborn rats. Dean & McCance (1947b) also found that the excretion of
diodone by infants was poor by adult standards, indicating that the
excretory powers of the tubules were also relatively underdeveloped at
birth (see also Williamson & Hiatt, 1947). These functions were found to
mature at different rates during the first year of life (Smith & Chasis,

1948; Rubin, Bruck & Rapoport, 1949; Vesterdal & Tudvad 1949; Falk & Benjamin, 1951; McCance, Naylor & Widdowson, 1954; Falk, 1955; Schnieden, 1957).

Progress in the study of neonatal renal function now became extremely rapid, with workers from both sides of the Atlantic participating in it (see McCance 1948, 1950; and Smith, 1945, 1959). Threshold and non-threshold substances were studied and Dean & McCance (1948) confirmed that in adults the excretion of phosphates showed signs of a diurnal rhythm and that the phosphate clearances of infants were always well below those of adults. Even before this the urine of neonatal children and other animals had been shown to contain very small amounts of phosphates, and this absence of the usual mammalian buffer substances was inevitably followed by very low titratable acidity (Gordon, McNamara & Benjamin, 1948; Rubin, Calcagno, Rubin & Weintraub, (1956). McCance & Hatemi (1961; and Hatemi & McCance, 1961a, b) studied the response of infants to an acid load in the form of ammonium chloride or calcium chloride, both of which gave closely similar results. The clinical risks of $CaCl_2$ had already been revealed by Darrow, da Silva & Stevenson (1945). It was found that normal breast-fed infants seven days old (a) raised the $[H^+] \times 10^{-7}$ less rapidly and completely than adults, (b) excreted less of the H^+ generated by the drugs as titratable acidity, and (c) excreted almost as much ammonia/kg body weight as adults. The end results were clear, if not all the reasons for them. They were, moreover, similar in all the newborns studied, although there were minor differences between one species and another (Cort & McCance, 1954; McCance, 1960; Hatemi & McCance, 1961a, b).

By the 1950s the study of neonatal renal function had broadened out in several directions. The first of these was into comparative physiology. Species differences began to emerge; between puppies, pigs and rabbits on the one hand (Forrest & Stanier, 1966) and pigs, babies and puppies on the other in the way their newborns excreted water and salt (McCance & Widdowson, 1958a, b). Dicker & Heller (1945, 1951) also found interesting comparisons and contrasts between rats, humans and guinea-pigs. The differences between the pig and the baby have turned out to be in more than the kidney, for the urine from the bladder of the foetal pig appears to be hypotonic to the fluid obtained simultaneously from the pelvic cavity of the kidney where the sodium is probably reabsorbed (Stanier, 1971; and see also France, Stanier & Wooding, 1974).

Merely by following changes in the constituents of the blood and the urine of newborn animals valuable information began to come to light about their metabolism and renal function. The concentration of urea in the plasma of rats rose to a peak twenty-four hours after birth and then subsided. The concentration in infants did not reach its peak till the third day: that in piglets varied from litter to litter for reasons which were not clear, but only two litters of pigs were studied at that time (McCance & Widdowson, 1947; McCance & Otley, 1951; Joppich & Wolf, 1958). A re-examination of the question by Zweymuller, Widdowson & McCance (1959) corrected the impression given by Needham (1931) that the placenta is freely permeable to creatinine and urea by showing that the concentration of creatinine in the serum of piglets at birth usually exceeded that of the sow, indicating considerable impermeability of the placenta to creatinine. Observations on the plasmas of foetal lambs made by Alexander, Nixon, Widdas & Wohlzogen (1958a) had been similar. The concentration of creatinine in Zweymuller *et al.*'s piglets fell rapidly in the first three days after birth, whereas the concentration of urea invariably rose. Boylan, Colbourn & McCance (1958) provided evidence of similar changes in the sera of newborn guinea-pigs, which are more mature at birth. Alexander *et al.* (1958b) also found that the volumes of urine passed by the foetal lambs were greatest at or about 117 days' gestation, but the percentage of the filtrate reabsorbed rose towards term and with this the volumes of urine fell off, but were always large and the urine hypotonic. These observations on the volumes passed by foetal lambs were followed by those of Perry & Stanier (1962) in pigs.

The failure of the kidneys of the newborn to handle electrolytes and urea like those of adults was really much greater than the facts so far quoted revealed, and for this reason. A study of the volumes and nitrogen partition of the urine soon after birth (Thomson, 1944; Barlow & McCance, 1948) and of the metabolic rate over the first three days, showed that much less protein was being broken down by male infants without food in providing for the energy requirements of the body (0·075 g/kg body weight per 24 h) than by healthy young adults (0·17 g/kg body weight per 24 h) or elderly males (0·12 g/kg body weight per 24 h) (McCance & Strangeways, 1954).

How the kidney of the newborn maintains internal stability has gradually come to light. In adults there is no complication due to growth, and variations in the food make so little difference to the serum 'chemistry' that they can usually be neglected. A much wider view of these processes must be taken in the newborn, and at least two other factors must be taken into account in considering how the stability of the internal environment is maintained. The first of these is the capacity of the animal to grow – the anabolic tendency to which reference has just been made – and the second is the composition of the food. This in nature is usually so perfect that practically everything in it can be incorporated into new tissues, leaving very little to be excreted by the kidney. This holds for protozoa as well as pigs or human infants, for when gelatine, or even gelatine enriched with tryptophan, were the only sources of nitrogen supplied to a colony of *Tetrahymena pyriformis*, much larger amounts of nitrogenous end-products appeared in the medium than when casein was supplied (Rosen & Fernell, 1956). So important is the composition of the diet, particularly in the case of a fast-growing animal, e.g. a marsupial (Bentley & Shield, 1962) or the rat which appears to have such rudimentary renal function at, and soon after, birth that it should always be in the forefront of one's mind when one is attempting to rear an infant or a newborn animal on a diet not provided for it by nature (McCance & Widdowson, 1956, 1957; Wilkinson, 1973).

By the end of the 1940s most responsible people must have realized that experiments were being carried out on adults and on infants which would probably not have been accepted as ethical by Claude Bernard. Risks were being taken, small ones admittedly, over cardiac catheterizations, liver biopsies, interferences with normal newborn infants and, later, administration of radioactive tracers (Black, Davies, Emery & Wade, 1956), which were acceptable to different degrees by different people and in different countries. The position of the investigator had never been defined. This led McCance (1951) to make a considered statement on the subject, which he was advised not to publish. McCance revised his views to some extent (1959a) but meanwhile others had written in the same vein (Fox, 1954; Hungerland, 1958). Investigations on normal persons and on patients went on as before, each investigator doing what he thought was right in his own eyes. This was as it should be, provided that everyone behaved in a responsible way, now that legal and professional opinion had been alerted (Ladimer, 1955; Davidson, 1957; Ladimer & Newman, 1963), because publicity in itself is a good protection (Whalan, 1975). During the 1940s, 50s and 60s, moreover, great progress in infant physiology and its application to treatment was made both in Britain and the United States (Smith, 1945, 1959).

Perinatal physiology owes a great debt to a few people; Huggett was one, and his influence on the Physiological Department at St Mary's Hospital shows it (see, for example, McCarthy, 1934). He introduced Joseph Barcroft to the experimental possibilities of the ruminant and so set the stage for all the work which was subsequently to go on at Cambridge. Don Barron, moreover, another of the pioneers, was in collaboration with Sir Joseph at Cambridge, and also with Meschia and others at Yale, helping to open up the school there, which has contributed so much since that time. The war was a stimulus rather than the reverse. It focussed the attention of investigators on service requirements while it lasted, but it also widened their interests and introduced them to technical advances which they could apply to their problems when peace returned.

The fifth International Congress of Paediatrics in New York soon after the war brought everyone together and acted as a spark. The importance of Cicily Williams's work in Africa was becoming recognized and people all over the world suddenly woke up to the importance of perinatal nutrition (Smith, 1947; Dean, 1951). Some of the gloomy predictions about the subsequent effects since made, however, have not been fulfilled (Stein, Susser, Saenger & Marolla, 1975).

It was about 1947–1950 too that the rearing of premature infants became such a challenge. The clinicians had the field to themselves for a time since the human infant was the only known mammal to survive after premature delivery. A series of conferences were staged some years later by the Macy Foundation (Lanman, 1956–1960) to try to solve some of the difficulties. The clinicians tackled the problem largely by trial and error, and their failures as well as their successes made great contributions to knowledge. Premature infants, for example, were often starved for the first four or more days after their birth (Hansen & Smith, 1953) and there was a vogue at one time for rearing them at temperatures as low as 36 °C. Although high pressures of oxygen had long been known to be toxic (Bean, 1945), prematures were for a time reared in an atmosphere of oxygen when they were blue with respiratory distress. This led to the discovery of a hitherto unrecognized lesion of the eye in infants termed retrolental fibroplasia (Hipsley, 1952; Hey, 1973).

A glance at the report of the Cold Spring Harbor Symposium held in 1954 is an indication of the interests and persons represented there. Lind and Wegelius in Sweden and an Oxford group, motivated by Geoffrey Dawes, were making spectacular advances on how the circulation was rerouted and the supply of oxygen to all parts of the body altered after a successful birth. The recent paper by de Swiet, Fancourt & Peto (1975) on the systolic blood pressure of normal human infants over the first week of life, as determined by an ultra-sound technique, and the one by Davis, Dewar, Tynar & Ward (1975) on the immaturity of the papillary muscles of the cat's right ventricle, show that the subject is still alive, and fundamental clinical information has been added in the interval. The possibilities of endocrine involvement in foetal welfare was brought out at the same symposium by Parkes and by Jost. Foetal and adult haemoglobin came in for attention, and Kenneth Cross described a new apparatus to study the respiratory movements of newborn infants asleep or awake. This he was to improve and work with for years, first at St Mary's and later at the London Hospital (Cross, 1949; Cross & Oppé, 1951; Cross & Warner, 1951; Cross, Hooper & Oppé, 1953). The resistance of the neonatal animal to anoxia was discussed by Miller (Fazekas, Alexander & Himwich, 1941; Adolph, 1948), and Brambell held the audience with an account of the passage of antibodies from the mother to her offspring. The Symposium was a milestone.

In 1959 Meschia, Wolkoff & Barron introduced a technique at Yale for studying aspects of foetal development in unstressed sheep by working with sterile catheters allowed to remain *in situ* for some time. For later examples of such work see Meschia, Cotter, Breathnach & Barron, 1965, and Crenshaw *et al.*, 1968. This technique has been much elaborated, and animals, particularly sheep, have now been maintained for weeks with in-dwelling catheters in selected vessels, vesicles and body fluids (see also Crenshaw *et al.*, 1973).

The founding of the Neonatal and Blair Bell Societies between 1959 and 1962 acted as a stimulus in this country, and the *British Medical Bulletin* (1961) devoted a number to foetal and neonatal physiology which was edited by Kenneth Cross. Many of the topics which had been aired at the Cold Spring Harbor Symposium came up again, and new ones, such as the reaction of the newborn animal to its environmental temperature. It was about this time that the physiologists were coming to the rescue of the clinicians over this and making important contributions to human and animal physiology at the same time (Brück, 1959; Hill & Rahimtulla, 1965; Varga, 1959; Alexander, 1961; Moore & Underwood, 1963; Silverman, Fertig & Berger, 1958; Silverman, Agate & Fertig, 1963; Hahn *et al.*, 1963; Mestyan, Jarai, Bata & Fekele, 1964; Mount, 1963, 1964; and see McCance, 1959*b*). The article by Davis and Tizard in the *British Medical Bulletin* (1961) on the comparative approach, considering both infants and animals, makes interesting reading in the light of later work in the paediatric units. Soon after this, small helpful conferences were held at the Ciba Foundation (Wolstenholme & O'Connor, 1961) and in Holland (Jonxis, Visser & Troelstra, 1964). At the former the respiratory distress syndrome received some attention and during the latter the role of brown fat in keeping the newborn rabbit warm without shivering was beautifully set out by Dawkins and Hull. In 1966 the *British Medical Bulletin* devoted a second number to the foetus and the newborn. Some of the subjects previously discussed were elaborated, and Reynolds and Strang discussed surfactant with due regard to the work of Pattle (1958, 1965) in previous years.

The possibility of keeping an immature foetus alive and fully functional in surroundings wholly foreign to it has stimulated thinkers for over 100 years (Walker & Danesh, 1973), for if such a thing could be achieved the physiological and therapeutic benefits would be enormous. The theoretical requirements can now be more or less defined and the technical equipment is obtainable. Of the more recent attempts those of the Swedes on foetuses obtained by legalized abortions were probably the first (Westin, Nyberg

& Enhörning, 1958), closely followed by those of groups in Canada (Callaghan & Angeles, 1961), at Cambridge (Lawn & McCance, 1962; Lawn, McCance & Thorn, 1967) and at St Mary's Medical School (Nixon, Britton & Alexander, 1963; Alexander, Britton & Nixon, 1964). The position was reviewed by Alexander, Britton & Nixon (1966), but progress has continued with human and other mammalian foetuses (Chamberlain, 1968); and five years later came the book edited by Austin (1973) which should be consulted for its technical and physiological interest.

Dawes (1968) wrote an excellent review of fields largely covered by the interests of his own department, but wider than volume **22** of the *British Medical Bulletin* (1966). Many spoke at the Barcroft Centenary Symposium (Comline, Cross, Dawes & Nathanielsz, 1973). Hathorn for instance (see also Hathorn, 1974, 1975), following the work of Prechtl & Lenard (1967), discussed the respiratory movements of infants during the two kinds of sleep.

Michael Purves has always been a disciple of Sir Joseph, and he and his colleagues have recently published from Bristol what would appear perhaps to be a major contribution to neonatal respiratory physiology. By taking simultaneous tracings of respiratory activity in the phrenic and external intercostal nerves, and from localized inspiratory and expiratory units in the medulla, and recording at the same time variations in tracheal pressure, they have reached a conclusion which would seem to the outsider to be wholly plausible, namely that there is no true onset of respiration at birth. The visible changes in breathing after delivery really form part of a sequence of events which can be traced back into the previous minutes or even hours (Ponte & Purves, 1973; Bystrzycka, Nail & Purves, 1975).

In 1975 a third number of the *British Medical Bulletin* devoted to perinatal research appeared. It contained some overlaps with the Barcroft Centenary Symposium and other conferences, the article by Hathorn, for example. Topics included an article by M. E. Avery (see Avery, 1975) dealing with the theoretical side of the respiratory distress syndrome and one by Reynolds (see Reynolds, 1975) on the more practical aspects (see also Gluck & Kulovich, 1973).

Joan Mott's contribution took us back to perinatal renal function. One could have wished that the erythropoietic problem could have been brought up to date as well as the renin angiotensin one (McCance, 1972), but this was done by Zanjani, Gidari, Peterson, Gordon and Wasserman at the Barcroft Centenary Symposium (Comline *et al.*, 1973). It is evident from the articles by Shelley, Barrett and Milner, Nathanielsz and Challis, and Thorburn that the development of the endocrine glands has become a big field in perinatal research, and that whether the foetus does or does not organize the time of its own delivery depends upon the maturity of the endocrine system as a whole (Comline & Silver, 1966; Ash *et al.*, 1973; Liggins, 1973).

It is perfectly clear from what has already been said that the study of a subject nowadays recognizes no boundaries. Anatomy, physiology, biochemistry and pharmacology are all drawn into the net. All of them demand specialized knowledge, yet the distinction between them for teaching remains and seems rather academic (Hopkins, 1938).

GROWTH – GENERAL PRINCIPLES AND CRITICAL PERIODS

Satisfactory growth is one of the best indications of health before maturity, and the word can be applied to physical or to mental growth. Growth in size involves anatomical changes in shape (Jackson, 1928), the deposition of more tissue, and, therefore, new tissue. New tissue, moreover, is never the same as that already in the body because the composition of the body, its chemistry in other words, changes with development. Biochemical development, therefore, goes hand in hand with physiological development and is even more finite and fundamental. There are now many books on the subject (e.g. Brozek, 1965; Timiras, 1972; Winick, 1972; Cheek, 1975) which no one person could review, and an author of any article on the subject must be very selective.

The first longitudinal study of the growth of a young man was made many years ago. It is one of the classics on human growth (see Scammon, 1927). The growth curve shows the long period of growth in the perinatal period and before puberty, and the so-called puberty growth spurt, both of which are characteristics of mankind. Other animals do not exhibit them and the growth curve of a rat or a pig is a smooth S-shaped one, although, by employing a suitable time-scale for the life-span of each animal, all the points can be combined into one curve after puberty (Brody, 1945). All growth curves extend far beyond the perinatal period, but the general pattern of growth at that time often helps to predict the final stature that an individual will attain (Tanner, 1962). Growth, therefore, should always be considered from its beginning to its end.

The increase in the weight of the body shows us only the integration of all the events which have been going on inside it. The percentage of water in the body, for example, falls with age from conception to about one year after birth (Widdowson, 1968), and it falls further as time goes on. For one reason, variable amounts of fat are deposited over the years. Table 1, which is adapted from Elsie Widdowson's (1968) paper, shows the amount and percentage of fat in a human male at various times during his life. Her rather hypothetical male adult contained 16% of fat, but a body has

TABLE 1. The total amounts of water, fat, protein, calcium, phosphorus, iron, copper and zinc in the body at various ages

Gestational age (weeks)	Weight (g)	Water (g)	Fat (g)	Protein N × 6·25 (g)	Ca (g)	P (g)	Fe (mg)	Cu (mg)	Zn (mg)
13	30	27	0·2	2·5	0·09	0·09	—	—	—
17	200	177	1·0	17·5	0·70	0·6	10	0·7	2·6
23	500	440	3·0	43·8	2·2	1·5	28	2·4	9·4
26	1 000	860	10·0	87·5	6·0	3·4	64	3·5	16·0
31	1 500	1 270	35·0	125·0	10·0	5·6	100	5·6	25·0
33	2 000	1 620	100·0	232·0	15·0	8·2	160	8·0	35·0
35	2 500	1 940	185·0	307·0	20·0	11·0	220	10·0	43·0
40	3 500	2 400	560·0	388·0	30·0	17·0	280	14·0	53·0
Adult	65 000	36 000	15 000·0	10 500·0	1 120·0	600·0	3 700	85·0	1 400·0

been analysed which contained only 4 % of fat, and a man who weighed 267 kg has been found to contain 70 %. Other features of growth, however, can be derived from this table which gives the total amount of water, protein, calcium, phosphorus, iron, copper and zinc in the body from the thirteenth week of gestation till term, and once more in the hypothetical adult. Some are worth a special note. The body consists of organs, and a number of these contribute a very large quota to the total amount of some of the inorganic elements in the body. The skeleton for example contributes about 98 % of the calcium, though not so much of the phosphorus, which is spread over the soft tissues as well. Most of the iron is circulating in the blood stream within the erythrocytes. The amount of copper in the liver goes up extremely fast during gestation and the liver contains about 60 % of the copper in the body at term, but this bizarre distribution does not persist into adult life. A similar sequence of events holds for the calf, guinea-pig and rabbit, and now it would seem for the lamb (Williams & Bremner, 1976). The amount of zinc in the body is large, relative to copper, and the liver is never the locus for much of it. By the time a man is mature, most of the zinc he contains is in his skeletal muscles. As with vitamin D – see p. 135) – the quantities of some of the trace elements required in the perinatal period are critical. The right amount is very valuable; too much or too little may be highly undesirable.

Even in the fat-free body the percentage of water continues to fall after birth till it reaches the adult figure of 72 %, and the percentage of protein to rise. These are generalizations which apply to all species. Figures for the pig were given by McCance & Widdowson (1954). This, however, is only the beginning of the complexities of growth. The increasing concentration of protein indicates cellular enlargement and the cells contain water, with potassium as their main electrolyte. Exceptionally, the erythrocytes in a few species contain sodium; those of dogs, cats and of some sheep are examples (Cohn & Cohn, 1939; Evans, Harris & Warren, 1958; Widdowson & Dickerson, 1964). The main electrolyte in the extracellular water of the body is sodium. Consequently with growth the percentages of extracellular fluid and the sodium in it are falling and being replaced by corresponding amounts of intracellular water, protein and potassium (Widdowson, 1968). The study of these electrolytes and water compartments of the body goes back a long way (Prout, 1831; O'Shaughnessy, 1832; Rogers, 1909), but the recognition of their importance in perinatal physiology and human therapy probably owes more to Ödön Kerpel-Fronius (1940, 1959), James Gamble's influence (1923, 1953, 1954) and Dan Darrow (1946; and see Darrow *et al.*, 1949) than most physiologists realize today (see also Nadal, Pedersen & Maddock, 1941; Elkinton & Danowski, 1955).

For many in recent years the study of growth has become one of cell division and cell enlargement. The spokesmen have been Cheek (1968), Winick and Dobbing. The political and emotional importance of the brain to man has led Winick and Dobbing to devote much or all of their attention to that organ. Davison & Dobbing (1961, 1966) first pointed out that all structures in the body were not subject to the rapid turnover of their constituent molecules that had been suggested by the first isotope studies, and that this was well demonstrated by the myelinated structures of the brain. Later they pointed out that any delay in the rate of growth due to undernutrition, always a point of great importance in the political arena, was most likely to injure the brain when myelination and cell division were going on most rapidly; for myelin was so stable that, if abnormally formed owing to the undernutrition, it would not be likely to break down and could never be replaced or repaired. As originally stated, this theory took no account of whether the time from conception when the rapid development was due to take place might have limited temporal boundaries, or whether the delay in cell multiplication and myelination might be made up later, even at a slower rate. Dobbing has modified these ideas in the light of his own work and that of others (Dobbing, 1974), but there are still difficulties to be overcome before all the work on this subject can be reconciled.

Winick & Noble (1965, 1966) emphasized an important principle of growth by showing that it consisted of two stages: (1) nuclear, i.e. cell division, and (2) cell enlargement. The former could only take place up to a certain time from conception, particularly in certain organs, and it could be delayed by undernutrition. If this delay prevented the full number of cells in an organ being formed before the closing date, so to speak, the organ remained small throughout the animal's life. Cells could enlarge, however, at any age, and therefore a period of undernutrition later in life merely reduced for a time the size of the cells – and so macroscopically of the organ (Winick, Brasel & Rosso, 1972).

The skeletal muscle cell is multinucleate and so one cannot get at the number of cells, their size and other measurable components in any named muscle, as one can in the lung: namely by measuring the total amount of DNA in a sample of it and dividing by the amount of DNA present in each diploid nucleus of the species under investigation (Winick *et al.*, 1972). Enesco & Puddy (1964), MacConnachie, Enesco & Leblond (1964), Widdowson (1968), Widdowson, Crabb & Milner (1972) and Cheek (1975) have devoted considerable time and thought to the problems. The total DNA in human muscle increases by about ten times in girls and fifteen times in boys between birth and maturity. At the same time the protein/ DNA ratio has about doubled and the amount of protein by twenty to thirty times (Cheek, 1968). There are further complications about skeletal muscle because some of the nuclei present in any named muscle lie out-side the muscle cells and belong to the connective tissue. Perhaps more progress has been made in the biochemistry and structure of heart muscle before and after birth (see Table 5 in Widdowson, 1968), but Cheek has made some interesting calculations about the nuclei and growth of skeletal muscle in male and female rats at puberty (Cheek, Brasel & Graystone, 1968).

When Widdowson & McCance (1960) used the technique of Kennedy (1957) of suckling newborn rats in large and small groups, and showed some of the simpler results that could be got out of it, they entitled the paper 'Some effects of accelerating growth'. They might equally well have called the paper 'Some effects of retarding growth' for what they had been doing was accelerating and retarding the rate of growth of rats around a mean which had hitherto been accepted as the normal one. The spirit of the times, however, was that one can not have too much of a good thing – *vide ante* p. 135 – and the general interpretation of the experiments was that the large rats were the normal ones and the small ones were undernourished. Widdowson and McCance were careful never to say so, for they realized they were on dangerous ground and statements in the paper show this.

Owing to the immaturity of the rat at birth these experiments were in fact the first ones to study the effect of varying the nutritional intake of an animal before the critical period in its life at which the hypothalamic centres take over the regulation of the animal's appetite and subsequent way of life. There are equally critical periods in a rat's sexual life (Barra-clough, 1961; Harris & Levine, 1962). Because of this, the small animals showed none of the physiological signs of undernutrition. Their appetites were integrated with their size and they ate accordingly. They were not lean and overhungry. They showed no sign of 'catch up' growth even in the presence of unlimited food, and their absolute weight tended to fall further and further behind that of the large rats. They behaved as though they were genetically small, slow-growing but perfectly proportioned little animals. Their brains and other organs were smaller than those of the large rats at each chronological age. They contained fewer cells, and where their behaviour was regulated by size they showed the expected signs of immaturity and lack of experience. Puberty was delayed for example and all that goes with it. Widdowson & McCance (1963) hinted that results similar to those in rats would only be achieved in guinea-pigs or pigs, and probably also in man, if the deprivation of food took place before birth. This has now been demonstrated experimentally (Widdowson, 1971), and explanations put forward (Widdowson & McCance, 1975).

It is difficult to know what to make of the theories of Winick or Dobbing in the light of these findings. Were they studying undernourished animals at all? Have the results got any bearing on protein deficiencies and energy deficiencies, or the weaning diseases that so preoccupy thinking about growth in the backward communities?

All experiments on undernutrition previous to those of Widdowson & McCance (1960) commenced after the critical period. Growth can be delayed so long, however, in a suitable species by this technique that it is difficult to see how cells can never multiply late in life. The famous experiments on rats by McCay, Maynard, Sperling & Barnes (1939) began after weaning, those on cockerels by Lister, Cowen & McCance (1966) when the birds weighed about 90 g, and those of Lister & McCance (1967) on pigs not until ten days after birth. In all these experiments growth in weight can be restricted for times equal or over the normal growth period of the animal, and even for longer than the normal life-span of the animal, and it will still show 'catch up' growth at the end of it. Its extent, however, becomes more and more limited as time goes on, particularly in males, but this is probably a matter of ageing and not perinatal physiology (McCance, 1968; McCance & Widdowson, 1974; Widdowson & McCance, 1975).

The truth is that undernutrition, or any other tactic that delays development, delays all aspects of it whatever the species or its rate of growth, even those which appear to be most closely linked to chronological age and to be independent of the plane of nutrition and increase of mass, such as the opening of the eyes and the appearance of the teeth (Widdowson & McCance, 1960; Tonge & McCance, 1973). Many have not appreciated this. What is not so sure, moreover, is whether the growth and development of some of these parts is so closely linked to chronological time that they must be completed within a limited period of it – or not at all – and this is the important issue where growth is concerned (Dobbing, 1974).

Although it undoubtedly occurs after birth in rats, the critical period may be assumed to occur before birth in guinea-pigs, farm animals and man, and the effect of varying the plane of nutrition of pure-bred pregnant guinea-pigs on the size and rate of growth of the newborn has been mentioned. It has not so far been found possible to exploit this commercially in farm animals. Sheep, for example, are usually kept and carried through pregnancy economically rather than scientifically, but a spell of good feeding in late pregnancy is known to improve the size and growth of the newborn lambs (Hammond, 1956).

The diets of some women during pregnancy may be very faulty, but semi-starvation *per se* reduces but slightly a baby's weight (Smith, 1947) for the mass of a whole conceptus is only quite a small fraction of that of the mother. The mass of a baby depends upon the height of its mother nevertheless (Thomson, 1968), and genetics enters into its subsequent growth and development as it does in the crossbred stains of animals (Walton & Hammond, 1938). Infants, however, do tend to follow the percentile line on which they are born if they are healthy and satisfactorily fed.

Infants may be small at birth because they are born prematurely without any check on their development *in utero*, or because their growth has been interfered with for some reason which has not led to premature delivery, so that they are too small for their gestational age. The former have no parallel among the lower mammals, the latter have many of the features of the badly placed foetal mouse (McLaren & Michie, 1960; McLaren, 1965), or the runt pig (Widdowson, 1971).

The problem of feeding the very small premature infant is a very real one and has been set out by Shaw (1973, 1974). The nub of the trouble may be presented in a few words. Although these children can often be reared by modern techniques their progress in g/day compared with growth *in utero*, whatever the milk provided for them, is often poor. The rate of accumulation of calcium *in utero*, moreover, is likely to exceed the amount contained in the breast milk or SMA they can ingest, even if it is given with great care (Valman, Heath & Brown, 1972).

BIOCHEMICAL PREPARATIONS FOR BIRTH

The accumulation of glycogen in the body during gestation has been a matter of interest since the first half of the last century. It has been described in recent years by Shelley (1960, 1962, 1964), Zetterström (1961), Dawes, Mott, Shelley & Stafford (1963), Shelley & Neligan (1966) and by Shelley, Bassett & Milner, (1975), and the observations have been similar in most species, although the amounts found in the various sites differ. The glycogen is a protection against anoxaemia as well as starvation. It builds up in the lungs and also seemingly in the heart till mid-gestation, but in the liver and muscles till birth at, or before, term. It then falls in a matter of hours to very low levels in the liver and more slowly in the muscles, particularly in premature infants, and symptoms and signs of hypoglycaemia may ensue. The newborn of all species tolerate hypoglycaemia much better than adults, but levels of 20 mg glucose /100 ml in infants are dangerous and treatment except by intravenous infusion may be unsatisfactory. The changes in the respiratory quotient are those to be expected as the infant turns over to its reserves of fat, but only the guinea-pig and the human infant contain appreciable amounts of fat, and here again the premature infant is at a disadvantage.

The newborn pig contains practically no fat but considerable amounts of glycogen in its liver and skeletal muscles. By drawing upon these and by gluconeogenesis it can maintain itself alive for 24 h or so even at quite low environmental temperatures. Its body temperature, however, will probably be falling, and normally a newborn pig weighing less than 1000 g depends upon food at frequent intervals for its day-to-day survival (McCance & Widdowson, 1959).

There are two aspects of fat metabolism which are of peculiar interest to perinatal physiologists. The first is the way in which brown fat participates in thermoregulation of the newborn and the second is the passage of fatty acids across the placenta. The study of non-shivering thermogenesis goes back a long way in cold-acclimated rats (Smith & Hoijer, 1962; Smith & Hock, 1963; Smith & Roberts, 1964), and since 1960 has covered arousal from hibernation (Smith & Horwitz, 1969). Heim & Hull (1966) and Hull (1974, 1975) have made a special study of the location and metabolism of brown fat in rabbits. Brown fat is only found in special sites. The main deposits of it are round the arteries of the neck and parts of the mediastinum, and in rabbits and some other species it forms a large interscapular pad. Histologically the adipocytes in brown fat are packed with mitochondria and the function of these cells is to produce heat if the surface of the body is exposed to cold. The fatty acids in brown adipocytes are similar to those in white ones, and the stimulus to the brown adipocytes, which causes them to break down the fat in them and to metabolize the acids, reaches them through the sympathetic-noradrenaline system. The fatty acid oxidation probably goes by the usual pathways in the adipocytes in brown fat. The unusual thing is that the heat of oxidation is set free immediately as heat, so some 'uncoupling' must go on as with dinitrophenol poisoning and with overproduction of thyroid hormones, and Hull has now suggested that large quantities of the fatty acids themselves can act as uncoupling agents.

By feeding pregnant guinea-pigs from mid-gestation (Pavey & Widdowson, 1975) on diets containing maize oil, beef tallow or the small amount of fat in the control diets, large differences were established between the amounts of fatty acids, particularly C18:2 acids (as g/100 g of total fatty acids) in the newborn young. Those of the mothers fed on maize oil contained much more of the 18:2 acids and this difference increased further after birth if the young continued to be fed on the oil.

We have known for some time that the fatty acids in the milk formula used for Dutch children after birth consisted of 58 % of the C18:2 acids – as g/100 g of total fatty acids. It was also known that the consumption of a 'soft' margarine has become very usual at all ages in Holland. This led Widdowson, Dauncey, Gairdner, Jonxis & Pelikan-Filipkova (1975) to study the amounts of C18:2 acids in the body fat (as g/100 g of total fatty acids) of Dutch children, and of British children fed on cow's milk. They found the British children to have 1·0 % of the C18:2 acids in their body fat at birth as against the Dutch babies 2·9 % and these differences were highly significant. The difference increased greatly, moreover, after birth. The serum cholesterols were lower, however, in the Dutch babies than those found in the sera of British children, whether the latter had been fed on breast milk or cow's milk. The implications of these results may be considerable.

BIOCHEMICAL DEVELOPMENT

Behind much of the perinatal physiology already described lies a world of biochemistry of the immature and developing organism (Driscoll & Hsia, 1958; Dawkins, 1959; Benson, 1971). The maturation of the system which converts substances such as bilirubin to the diglucuronide through the uridine diphosphoglucuronic acid system (Brown, 1968) is a simple example. There is no call for this mechanism till the intestinal tract begins to function as an excretory organ after birth, but it may not mature rapidly enough then to prevent an undesirable rise in the form of bilirubin in the serum which gives the 'indirect' van den Bergh response.

Other substances are, however, also made water soluble and excreted by this system: bromsulphthalein is, for example (Mollison & Cutbush, 1949; Yudkin, Gellis & Lappen, 1949), and more importantly nowadays phenobarbitone is. It has been noted that the children of mothers on phenobarbitone often have very low bilirubin concentrations in their sera, suggesting that the enzyme system was actively induced by the drug.

The potential benefit of rapid development of glucuronyl transferase activity in early extrauterine development, not only for the prevention of neonatal jaundice, but for the metabolism of drugs and hormones is obvious. Similar advantage to the foetus is not obvious, and should be approached with some caution. Foetal ability to conjugate might, in the case of bilirubin at least, prevent foetal clearance of bilirubin via its normal intrauterine route through the placenta.　　　　(Brown, 1968.)

Back in 1958 it was noted that the bilirubin in the serum of infants, and particularly premature infants, was bleached by light. This was followed by further observations and comments on both sides of the Atlantic (Lucey, Ferriero & Hewith, 1968; Behrman & Hsia, 1969; Editorial, 1970*a*, *b*, 1972). Destruction of the bilirubin appears to be by a process of oxidation: biliverdin, and other ill-defined, but probably non-toxic, water-soluble substances are formed. The light itself has to be carefully applied and the treatment has been little used except by a few people for prematures.

THE SPECIAL SENSES

There have been developments recently in the physiology of the special senses, which it seems just possible to touch upon in an article of this kind as it is all very pertinent to perinatal physiology. I refer to the subject of pheromones.

The ability of dogs from all over the countryside to collect round a bitch on heat must have been an observation made centuries ago, and is a good example of the action of a volatile pheromone. Dogs, however, are not the only creatures with a highly developed sense of smell. Insects in particular have one (Loher, 1960–1961); yet some insect pheromones are not volatile. The 'Queen Bee' substance is not, for instance (Butler, Callow & Johnston, 1961–1962), and there are suggestions of others.

Hilda Bruce (1959, 1960) noted that a recently mated female mouse failed to produce young if housed with or near a strange male. The mated female was not upset by her own stud male or by strange males if they had been castrated. Bruce & Parrott (1960) immediately showed that removing the olfactory bulb from the mated females abolished the inhibition and pregnancy ensued normally. It was subsequently shown by Dominic (1966*a*) that the urine of the strange male was the source of the block to pregnancy, and (Dominic, 1966*b*) that the failure of the fertilized ovum to implant was caused by defective prolactin secretion. Injecting prolactin removed the block and so did an ectopic pituitary implant, for both encouraged the corpora lutea of pregnancy to mature normally. It had been their failure to do so which led to the failure of the fertilized ova to imbed.

Cowley and others have been removing the olfactory bulbs from mice at various times after birth. A highly complicated situation is immediately created as it also seems to be in young rats (Kling, 1964). Events seem to depend upon (*a*) the size and placement of the lesions, (*b*) the post-natal age of the mice at which the lesions are made – one of the critical periods of growth may be coming in here, and (*c*) the presence and number of the remaining littermates.

If neonatally bulbectomized, the mice may fail to seek the mother, become very active, wander away from the warmth of the nest and never return. If the operation is not carried out till the fifth day of life the animals may become less active than littermates, and remain so throughout the whole growth period (Cowley, personal communication). A situation as complicated as this defies analysis in a few words and the original papers will have to be consulted (Bruce, 1970; Cowley & Wise, 1970, 1972; McClelland & Cowley, 1972; Cooper & Cowley, 1976; Cowley, 1976).

IMMUNOLOGICAL PROVISION FOR THE NEWBORN

If the foetus is now to be regarded as a 'tolerated allograft' (Voisin & Chaovat, 1974), the old conception that the foetus and the mother lived in untroubled symbiosis will have to go (Smith, 1968) and be replaced by the thought that they are really rubbing along together in a state of armed neutrality (Faulk *et al.*, 1974). Nevertheless, if the newborn creature is to survive all the microbial hazards of its new environment it must already have, or be able to obtain almost at once from its mother, the immune bodies required to tide it over till it can make its own. Each species seems to have evolved its own method of passing them over and in cattle, pigs and kittens the process was studied histologically by Comline, Roberts & Titchen (1951) and Comline, Pomeroy & Titchen (1953). Brambell (1958) covered the subject fairly fully and summarized his views again at a meeting in London (1961) and earlier in the United States as already described (see also Bangham, 1961). Papers continued to emerge from his A.R.C. Unit in Bangor (Brambell, Halliday & Hemmings, 1960, 1961; Morris, 1965–1966) and later from the A.R.C. Unit at Babraham (Pierce & Smith, 1967) and the Zoological Department at Nottingham (Morris, 1975). Miller (1966) gave a brief but very lucid account of the state of knowledge at that time which will probably be sufficient for most conventional physiologists. The jargon has become very forbidding, and one wonders how necessary it all is!

THE FUTURE

The 1960s opened without any outward and visible changes having taken place in the old doctor–patient relationships. By about 1965, however, the social services, technology, the media, and the much advertised rights of the individual had so increased the sophistication of the patients that many of them began to feel they knew as much as their doctors and to challenge their decisions, to institute litigation, etc. for which they knew they would rarely have to pay.

At the same time certain investigators had disregarded the warnings, exhortations and precautions of their predecessors. A potentially dangerous situation was already developing when Pappworth's (1967) provocative book appeared. The word 'research' was vilified and this had unfortunate consequences. Ethical committees were set up by the powers that be which have made it almost impossible to touch a human infant, or foetus for that matter (Porter, 1973). The honest investigator was no longer allowed to appraise the problem before him and make his own judgement upon it, but a few people have retained the courage of their convictions. 'I still feel that the best possible control of all in this matter of ethics and morality must be the conscience of the individual investigator' (Whalan, 1975; Rhodes, 1975a).

Politics cannot be kept out of medicine nowadays either in its practice or research (Rhodes, 1975b) and things may take a very long time to come right. Meantime, progress is bound to slow down, unless of course, as has happened so often in this country, mutually satisfactory catch phrases, conventions and compromises such as 'informed consent' free the human physiologists from the shackles of a politically minded and usually uncooperative public.

REFERENCES

ADOLPH, E. F. (1948). Tolerance to cold and anoxia in infant rats. *Am. J. Physiol.* **155**, 366–377.
ALEXANDER, D. P., BRITTON, H. G. & NIXON, D. A. (1964). Survival of the foetal sheep at term following short periods of perfusion through the umbilical vessels. *J. Physiol.* **175**, 113–124.
ALEXANDER, D. P., BRITTON, H. G. & NIXON, D. A. (1966). Maintenance of the isolated foetus. *Br. med. Bull.* **22**, 9–12.
ALEXANDER, D. P., NIXON, D. A., WIDDAS, W. F. & WOHLZOGEN, F. X. (1958a). Gestational variations in the composition of the foetal fluids and foetal urine in the sheep. *J. Physiol.* **140**, 1–13.
ALEXANDER, D. P., NIXON, D. A., WIDDAS, W. F. & WOHLZOGEN, F. X. (1958b). Renal function in the sheep foetus. *J. Physiol.* **140**, 14–22.
ALEXANDER, G. (1961). Temperature regulation in the newborn lamb. 3. The effect of environmental temperature on metabolic rate, body temperatures and respiratory quotient. *Aust. J. agric. Sci.* **12**, 1152–1174.
ASH, R. W., CHALLIS, J. R. G., HARRISON, F. A., HEAP, R. B., ILLINGWORTH, D. V., PERRY, J. S. & POYSER, N. L. (1973). Hormonal control in pregnancy and parturition: a comparative analysis. In *Barcroft Centenary Symposium*, ed. COMLINE, R. S., CROSS, K. W., DAWES, G. S. & NATHANIELSZ, P. W., pp. 551–561. London: Cambridge University Press.
AUSTIN, C. R. (ed.) (1973). *The Mammalian Foetus In Vitro*. London: Chapman & Hall.
AUSTIN, J. H., STILLMAN, E. & VAN SLYKE, D. D. (1921). Factors governing the excretion rate of urea. *J. biol. Chem.* **46**, 91–112.
AVERY, M. E. (1975). Pharmacological approaches to the acceleration of fetal lung maturation. *Br. med. Bull.* **31**, 13–17.

BACON, J. S. D. & BELL, D. J. (1948). Fructose and glucose in the blood of the foetal sheep. *Biochem. J.* **42**, 397–405.

BAKWIN, H. (1937). Pathogenesis of tetany of newborn. *Am. J. Dis. Child.* **54**, 1211–1216.

BANGHAM, D. R. (1961). The exchange of serum proteins between mother and foetus in the rhesus monkey. *Proc. R. Soc. Med.* **54**, 993–996.

BARCROFT, J. (1925, 1928). *The respiratory Functions of the Blood:* Vol. 1 (1925) *Lessons from High Altitudes;* Vol. 2 (1928) *Haemoglobin.* Cambridge: Cambridge University Press.

BARCROFT, J. (1938). *Features in the Architecture of Physiological Function.* Cambridge: Cambridge University Press.

BARCROFT, J., COOKE, A., HARTRIDGE, H., PARSONS, T. R. & PARSONS, W. (1919–1920). The flow of oxygen through the pulmonary epithelium. *J. Physiol.* **53**, 450–472.

BARKLAY, H., HAAS, P., HUGGETT, A. StG., KING, G. & ROWLEY, D. (1949). The sugar of the foetal blood, the amniotic and allantoic fluids. *J. Physiol.* **109**, 98–102.

BARLOW, A. & McCANCE, R. A. (1948). The nitrogen partition in newborn infants' urine. *Arch. Dis. Childh.* **23**, 225–230.

BARNETT, H. L. (1940). Renal physiology in infants and children. 1. Method for estimation of glomerular filtration rate. *Proc. Soc. exp. Biol. Med.* **44**, 654–658.

BARNETT, H. L., HARE, K., McNAMARA, H. & HARE, R. (1948a). Measurement of glomerular filtration rate in premature infants. *J. clin. Invest.* **27**, 691–699.

BARNETT, H. L., HARE, W. K., McNAMARA, H. & HARE, R. S. (1948b). Influence of post-natal age on kidney function of premature infants. *Proc. Soc. exp. Biol. Med.* **69**, 55–57.

BARNETT, H. L., McNAMARA, H., HARE, R. S. & HARE, K. (1948). Inulin, urea, mannitol and PAH clearance ratios in premature infants. *Fedn Proc.* **7**, 5–6.

BARNETT, H. L., PERLEY, A. M. & McGINNIS, H. G. (1942). Renal physiology in infants and children. 2. Inulin clearances in newborn infant with extrophy of bladder. *Proc. Soc. exp. Biol. Med.* **49**, 90–93.

BARRACLOUGH, C. A. (1961). Production of anovulatory, sterile rats by single injections of testosterone propionate. *Endocrinology* **68**, 62–67.

BEAN, J. W. (1945). Effects of oxygen at increased pressure. *Physiol. Rev.* **25**, 1–147.

BEHRMAN, R. E. & HSIA, D. Y. Y. (1969). Summary of a symposium on phototherapy for hyperbilirubinemia. *J. Pediat.* **75**, 718–726.

BENSON, P. (1971). The biochemistry of development. *Clinics in Developmental Medicine*, Vol. **37**. London: Spastics International Medical Publications and Heinemann Medical Books Ltd.

BENTLEY, P. J. & SHIELD, J. W. (1962). Metabolism and kidney function in the pouch young of the macropod marsupial *Setonix brachyurus. J. Physiol.* **164**, 127–137.

BERNARD, C. (1855). *Leçons de physiologie expérimentale appliquée à la médecine faites au Collège de France.* Vingt-et-unième leçon, p. 397. Paris: Baillière.

BERNARD, C. (1965). *Introduction à l'étude de la Médicine Expérimentale.* Paris: Baillière and Son.

BERT, P. (1878). *La Pression Barometrique.* Paris: Martinet.

BLACK, D. A. K., DAVIES, H. E. F., EMERY, E. W. & WADE, E. G. (1956). Renal handling of radioactive potassium in man. *Clin. Sci.* **15**, 277–283.

BOYLAN, J. W., COLBOURN, E. P. & McCANCE, R. A. (1958). Renal function in the foetal and newborn guinea-pig. *J. Physiol.* **141**, 323–331.

BOYLE, R. (1670). Of the phenomena afforded by a newly kittened kitling in the exhausted receiver. *Phil. Trans. R. Soc.* **5**, (62), 2017–2019.

BRAMBELL, F. W. R. (1958). The passive immunity of the young mammal. *Biol. Rev.* **33**, 488–531.

BRAMBELL, F. W. R. (1961). The transmission of antibodies from mother to foetus: problems. *Proc. R. Soc. Med.* **54**, 992–993.

BRAMBELL, F. W. R., HALLIDAY, R. & HEMMINGS, W. A. (1960, 1961). Changes in ^{131}I-labelled immune bovine γ-globulin during transmission to the circulation after oral administration to the young rat. *Proc. R. Soc.* B, **153**, 477–489.

BRINKMAN, R. & JONXIS, J. H. P. (1935). The occurrence of several kinds of haemoglobin in foetal blood. *J. Physiol.* **85**, 117–127.

BRINKMAN, R. & JONXIS, J. H. P. (1937). Alkaline resistance and spreading velocity of foetal and adult types of mammalian haemoglobin. *J. Physiol.* **88**, 162–166.

BRITISH MEDICAL BULLETIN (1961). **17**, 79–174. Foetal and neonatal physiology London: The British Council.

BRITISH MEDICAL BULLETIN (1966). **22**, 1–99. The foetus and the newborn: recent research. London: The British Council.

BRITISH MEDICAL BULLETIN (1975). **31**, 1–95. Perinatal research. London: The British Council.

BRODY, S. (1945). *Bioenergetics and Growth.* New York: Reinhold.

BROWN, A. K. (1968). Bilirubin metabolism in the developing liver. In *The Biology of Gestation*, vol. 2, ed. ASSALI, N. S., pp. 355–373. London and New York: Academic Press.

BROZEK, J. (ed.) (1965). *Human Body Composition: Approaches and Application.* London & New York: Pergamon.

BRUCE, H. M. (1959). An exteroceptive block to pregnancy in the mouse. *Nature, Lond.* **184**, 105.

BRUCE, H. M. (1960). A block to pregnancy in the mouse caused by proximity of strange males. *J. Reprod. Fert.* **1**, 96–103.

BRUCE, H. M. (1970). Pheromones. *Br. med. Bull.* **26**, 10–13.

BRUCE, H. M. & PARROTT, D. M. V. (1960). Role of olfactory sense in pregnancy block by strange males. *Science*, **131**, 1526.

BRÜCK, K. (1959). Die Temperaturregelung in den ersten Lebenstagung. In *Die physiologische Entwicklung des Kindes*, pp. 41–53. Marburg an der Lahn: Fredrich Linneweh.

BUTLER, C. G., CALLOW, R. K. & JOHNSTON, N. (1961–1962). The isolation and synthesis of queen substance, 9-oxodec-trans-2-enoic acid, a honey bee pheromone. *Proc. R. Soc.* B **155**, 417–432.

BYSTRZYCKA, E., NAIL, B. S. & PURVES, M. J. (1975). Central and peripheral neural respiratory activity in the mature sheep, foetus and newborn lamb. *Respir. Physiol.* **25**, 199–215.

CALLAGHAN, J. C. & ANGELES, J. D. (1961). Long term extracorporeal circulation in the development of an artificial placenta for respiratory distress of the newborn. *Surg. Forum* **12**, 215–217.

CHAMBERLAIN, G. (1968). An artificial placenta. *Am. J. Obst. Gynec.* **100**, 615–626.

CHEEK, D. B. (1968). *Human Growth: Body Composition, Cell Growth, Energy and Intelligence.* Philadelphia: Lea & Febiger. London: Henry Kimpton.

CHEEK, D. B. (1975). *Fetal and Postnatal Cellular Growth.* New York: Wiley.

CHEEK, D. B., BRASEL, J. A. & GRAYSTONE, J. E. (1968). Muscle cell growth in rodents: sex difference and the role of hormones. In *Human Growth*, ed. CHEEK, D. B., pp. 306–325. Philadelphia: Lea & Febiger.

CLARKE, R. W. & SMITH, H. W. (1932). Absorption and excretion of water and salts by the elasmobranch fishes. *J. cell. comp. Physiol.* **1**, 131–143.

COHN, W. E. & COHN, E. T. (1939). Permeability of red corpuscles of the dog to sodium ion. *Proc. Soc. exp. Biol. Med.* **41**, 445–449.

COHNSTEIN, J. (1884). Blutveränderungen während der Schwangerschaft. *Pflügers Arch ges. Physiol.* **34**, 233–237.

COHNSTEIN, J. & ZUNTZ, N. (1884). Untersuchungen über das Blut, den Kreislauf und die Athmung beim Säugethier-Foetus. *Pflügers Arch ges. Physiol.* **34**, 173–233.

Cold Spring Harbor Symposia for Quantitative Biology (1954). **19**. *The Mammalian Fetus: Physiological Aspects of Development.* New York: Long Island Biological Assn.

COMLINE, R. S., CROSS, K. W., DAWES, G. S. & NATHANIELSZ, P. W. (eds.) (1973). *Barcroft Centenary Symposium.* London: Cambridge University Press.

COMLINE, R. S., POMEROY, R. W. & TITCHEN, D. A. (1953). Histological changes in the intestine during colostrum absorption. *J. Physiol.* **122**, 6P.

COMLINE, R. S., ROBERTS, H. E. & TITCHEN, D. A. (1951). Histological changes in the epithelium of the small intestine during protein absorption in the newborn animal. *Nature, Lond.* **168**, 84–85.

COMLINE, R. S. & SILVER, M. (1966). Development of activity in the adrenal medulla of the foetus and newborn animal. *Br. med. Bull.* **22**, 16–20.

COMMITTEE (1956). Hypercalcaemia in infants and vitamin D. *Br. med. J.* **2**, 149.

COOPER, A. J. & COWLEY, J. J. (1976). Mother–infant interaction in mice bulbectomised early in life. *Physiol. Behav.* **16**, 453–459.

CORT, J. H. & McCANCE, R. A. (1954). The renal response of puppies to an acidosis. *J. Physiol.* **124**, 358–369.

COWLEY, J. J. (1976). Olfaction and the development of sexual behaviour. In *Biological Determinants of Sexual Behaviour*, ed. HUTCHISON, J. B. London: Wiley (in press).

COWLEY, J. J. & WISE, D. R. (1970). Pheromones, growth and behaviour. In *Ciba Foundation Study Group*, Vol. **35**, ed. PORTER, R. & BIRCH, J., pp. 144–170. London: Churchill.

COWLEY, J. J. & WISE, D. R. (1972). Some effects of mouse urine on neonatal growth and reproduction. *Anim. Behav.* **20**, 499–506.

CRENSHAW, C., CEPHALO, R., SCHOMBERG, D. W., CURET, L. B. & BARRON, D. H. (1973). Estimations of the umbilical uptake of glucose by foetal lamb. In *Barcroft Centenary Symposium*, ed. COMLINE, R. S., CROSS, K. W., DAWES, G. S. & NATHANIELSZ, P. W., pp. 298–305. London: Cambridge University Press.

CRENSHAW, C., HUCKABEE, W. E., CURET, L. B., MANN, L. & BARRON, D. H. (1968). A method for the estimation of the umbilical blood flow in unstressed sheep and goats with some results of its application. *Quart. J. exp. Physiol.* **53**, 65–75.

CROSS, K. W. (1949). The respiratory rate and ventilation in the newborn baby. *J. Physiol.* **109**, 459–474.

CROSS, K. W., HOOPER, J. M. D. & OPPÉ, T. E. (1953). The effect of inhalation of carbon dioxide in air on the respiration of the full term and premature infant. *J. Physiol.* **122**, 264–273.

CROSS, K. W. & OPPÉ, T. E. (1951). A comparison between adults, full-term and premature infants in their respiratory response to oxygen. *J. Physiol.* **115**, 17P.

CROSS, K. W. & WARNER, P. (1951). The effect of inhalation of high and low oxygen concentrations on the respiration of the newborn infant. *J. Physiol.* **114**, 283–295.

ĆURČIĆ, V. G. & ĆURČIĆ, B. (1974). Effect of vitamin D on serum cholesterol and arterial blood pressure in infants. *Nutr. Metab.* **18**, 57–61.

CUSHNY, A. R. (1917). *The Secretion of Urine.* London: Longmans, Green.

DARROW, D. C. (1946). Advances in the treatment of diarrhoea in infants. *Texas Rep. Biol. Med.* **5**, 29–54.

DARROW, D. C., PRATT, E. L., FLETT, J., GAMBLE, A. H. & WIESE, H. E. (1949). Disturbances of water and electrolytes in infantile diarrhoea. *Pediatrics* 3, 129–156.

DARROW, D. C., DA SILVA, M. M. & STEVENSON, S. S. (1945). Production of acidosis in premature infants by protein milk. *J. Pediat.* 27, 43–58.

DAVIDSON, M. (1957). *Medical Ethics: a Guide to Students and Practitioners.* London: Lloyd-Luke (Medical Books).

DAVIES, H. W., HALDANE, J. B. S. & KENNAWAY, E. L. (1920). Experiments on the regulation of the blood's alkalinity. *J. Physiol.* 54, 32–45.

DAVIES, P., DEWAR, J., TYNAN, M. & WARD, R. (1975). Post-natal developmental changes in the length tension relationship of cat papillary muscles. *J. Physiol.* 253, 95–102.

DAVISON, A. N. & DOBBING, J. (1961). Metabolic stability of body constituents. *Nature, Lond.* 191, 844–848.

DAVISON, A. N. & DOBBING, J. (1966). Myelination as a vulnerable period in brain development. *Br. med. Bull.* 22, 42–44.

DAWES, G. S. (1968). *Foetal and Neonatal Physiology.* Chicago: Year Book Medical Publishers.

DAWES, G. S., MOTT, J. C., SHELLEY, H. J. & STAFFORD, A. (1963). The prolongation of survival time in asphyxiated immature foetal lambs. *J. Physiol.* 168, 43–64.

DAWKINS, M. J. R. (1959). Respiratory enzymes in the liver of the newborn rat. *Proc. R. Soc.* B. 150, 284–298.

DEAN, R. F. A. (1951). The size of the baby at birth and the yield of breast milk. Studies of undernutrition, Wuppertal 1946–9. *Med. Res. Coun. Spec. Rep. Ser.* 275, 346–378.

DEAN, R. F. A. & McCANCE, R. A. (1974a). Response of newborn infants to hypertonic solutions of sodium chloride and of urea. *Nature, Lond.* 160, 904.

DEAN, R. F. A. & McCANCE, R. A. (1974b). Inulin, diodone, creatinine and urea clearances in newborn infants. *J. Physiol.* 106, 431–439.

DEAN, R. F. A. & McCANCE, R. A. (1949). The renal responses of infants and adults to the administration of hypertonic solutions of sodium chloride and urea. *J. Physiol.* 109, 81–97.

DE GRAAF, R. (1668, 1672). See JOCELYN, H. D. & SETCHELL, B. P. (1972). *Suppl. No.* 17 of the *J. Reprod. Fert.*

DICKER, S. E. & HELLER, H. (1945). The mechanism of water diuresis in normal rats and rabbits as analysed by inulin and diodone clearances. *J. Physiol.* 103, 449–460.

DICKER, S. E. & HELLER, H. (1951). The mechanism of water diuresis in adult and newborn guinea pigs. *J. Physiol.* 112, 149–155.

DOBBING, J. (1974). The later development of the brain and its vulnerability. In *Scientific Foundations of Paediatrics*, ed. DAVIS, J. A. & DOBBING, J., pp. 565–577. London: Heinemann Medical.

DOHRN (1867). Zur Kentniss des Harns des menschlichen Fötus und Neugeborenen. *Mschr. Geburtskunde und Frauenkrankheiten* 29, 105–134.

DOMINIC, C. J. (1966a). Observations on the reproductive pheromones of mice. 1. Source. *J. Reprod. Fert.* 11, 407–414.

DOMINIC, C. J. (1966b). Observations on the reproductive pheromones of mice. 2. Neuro-endocrine mechanisms involved in the olfactory block to pregnancy. *J. Reprod. Fert.* 11, 415–421.

DRISCOLL, S. G. & HSIA, D. Y. (1958). Development of enzyme systems during early development. *Pediatrics* 22, 785–845.

EDITORIAL (1970a). Phototherapy for neonatal jaundice. *Lancet* i, 825–826.

EDITORIAL (1970b). Blue light and jaundice. *Br. med. J.* 2, 5–6.

EDITORIAL (1972). Phototherapy in neonatal jaundice. *Br. med. J.* 2, 62–63.

ELKINTON, J. R. & DANOWSKI, T. S. (1955). *The Body Fluids.* Baltimore: Williams and Wilkins.

ENESCO, M. & PUDDY, D. (1964). Increase in the number of nuclei and weight in skeletal muscle of rats at various ages. *Am. J. Anat.* **114**, 235–244.

EVANS, J. V., HARRIS, H. & WARREN, F. L. (1958). The distribution of haemoglobin and blood potassium types in British breeds of sheep. *Proc. R. Soc. B*, **148**, 249–262.

FALK, G. (1955). Maturation of renal function in infant rats. *Am. J. Physiol.* **181**, 157–170.

FALK, G. & BENJAMIN, J. A. (1951). Observations on water diuresis and on ureteral peristalsis in an infant with extrophy of the bladder. *Surgery Gynec. Obstet.* **93**, 159–166.

FAULK, W. P., JEANNET, M., CREIGHTON, W. D., CARBONARA, A. & HAY, F. (1974). Studies of the human placenta. 2. Characterization of immunoglobulins on the trophoblastic basement membrane. *J. Reprod. Fert. Suppl.* **21**, 43–57.

FAZEKAS, J. F., ALEXANDER, F. A. D. & HIMWICH, H. E. (1941). Tolerance of the newborn to anoxia. *Am. J. Physiol.* **134**, 281–287.

FORREST, J. N. & STANIER, M. W. (1966). Kidney composition and renal concentration ability in young rabbits. *J. Physiol.* **187**, 1–4.

FOX, T. F. (1954). Conflict of loyalties. *Lancet* ii, 416–419.

FRANCE, V. M., STANIER, M. W. & WOODING, F. B. P. (1974). Structure and function in urinary bladder of foetal sheep. *J. Physiol.* **239**, 499–517.

GAMBLE, J. L. (1953). Early history of fluid replacement therapy. *Pediatrics* **11**, 554–567.

GAMBLE, J. L. (1954). *Chemical Anatomy, Physiology and Pathology of Extracellular Fluid. A lecture syllabus.* Cambridge, Mass.: Harvard University Press.

GAMBLE, J. L., ROSS, G. S. & TISDALL, F. F. (1923). The metabolism of fixed base during fasting. *J. biol. Chem.* **57**, 633–695.

GARDNER, L. I. (1952). Tetany and parathyroid hyperplasia in the newborn infant: Influence of dietary phosphate load. *Pediatrics* **9**, 534–543.

GARDNER, L. I., MACLACHLAN, E. A., PICK, W., TERRY, M. L. & BUTLER, A. M. (1950). Etiologic factors in tetany of newly born infants. *Pediatrics* **5**, 228–240.

GLUCK, L. & KULOVITCH, M. V. (1973). Surfactant production and foetal maturation. In *Barcroft Centenary Symposium*, ed. COMLINE, R. S., CROSS, K. W., DAVIES, G. S. & NATHANIELSZ, P. W., pp. 638–641. London: Cambridge University Press.

GOODWIN, R. F. W. (1956). Division of the common mammals into two groups according to the concentration of fructose in the blood of the foetus. *J. Physiol.* **132**, 146–156.

GORDON, H. H., MCNAMARA, H. & BENJAMIN, H. R. (1948). The response of young infants to ingestion of ammonium chloride. *Pediatrics* **2**, 290–302.

HAHN, O., KOLDOVSKY, J., KREČEK, J., MARTINEK, J. & VACEK, Z. (1963). Temperature adaptation during postnatal development. *Fedn Proc.* **22**, 824–827.

HALDANE, J. B. S. (1921). Experiments on the regulation of the blood's alkalinity. *J. Physiol.* **55**, 265–275.

HALDANE, J. B. S. (1925). The production of acidosis by ingestion of magnesium chloride and strontium chloride. *Biochem. J.* **19**, 249–250.

HALDANE, J. B. S. (1940). On being one's own rabbit. In *Possible Worlds*, pp. 105–116. London: Evergreen Books.

HALDANE, J. B. S., HILL, R. & LUCK, J. M. (1923). Calcium chloride acidosis. *J. Physiol.* **57**, 301–306.

HALDANE, J. S. & PRIESTLEY, J. G. (1935). *Respiration.* Oxford: Clarendon Press.

HAMMOND, J. (1956). *Farm Animals*, 2nd edn. London: Edward Arnold.

HANSEN, J. D. L. & SMITH, C. A. (1953). Effects of withdrawing fluid in the immediate postnatal period. *Pediatrics* **12**, 99–113.

HARRIS, G. W. & LEVINE, S. (1962). Sexual differentiation of the brain and its experimental control. *J. Physiol.* **163**, 42P–43P.

HATEMI, N. & McCANCE, R. A. (1961a). The response of piglets to ammonium chloride. *J. Physiol.* **157**, 603–610.

HATEMI, N. & McCANCE, R. A. (1961b). Renal aspects of acid base control. 3. Response to acidifying drugs. *Acta Paediat.* **50**, 603–616.

HATHORN, M. K. S. (1974). The rate and depth of breathing in newborn infants in different sleep states. *J. Physiol.* **243**, 101–113.

HATHORN, M. K. S. (1975). Analysis of the rhythm of infantile breathing. *Br. med. Bull.* **31**, 8–12.

HEIM, I. & HULL, D. (1966). The blood flow and oxygen consumption of brown adipose tissue in the newborn rabbit. *J. Physiol.* **186**, 42–55.

HELLER, H. (1944). The renal function of newborn infants. *J. Physiol.* **102**, 429–440.

HELLER, H. (1947). Antidiuretic hormone in pituitary glands of newborn rats. *J. Physiol.* **106**, 28–32.

HELLER, H. & ZAIMIS, E. J. (1949). The antidiuretic and oxytoxic hormones in the posterior pituitary glands of newborn infants and adults. *J. Physiol.* **109**, 162–169.

HEY, E. (1973). Care of the full-term human infant. In *The Mammalian Foetus In Vitro*, ed. AUSTIN, C. R., pp. 251–355. London: Chapman & Hall.

HILL, J. R. & RAHIMTULLA, K. H. (1965). Heat balance and the metabolic rate of newborn babies in relation to environmental temperature; and the effect of age and weight on basal metabolic rate. *J. Physiol.* **180**, 239–265.

HIPSLEY, E. H. (1952). The incidence of retrolental fibroplasia in premature infants in Sydney. *Med. J. Aust.* **1**, 473–474.

HÖBER, R. (1933). Über die Ausscheidung von Zuckern durch die isolierte Froschniere. *Pflügers Arch. ges. Physiol.* **233**, 181–198.

HOPKINS, F. G. (1938). Biological thought and chemical thought: a plea for unification. *Lancet* **i**, 1147–1150.

HUEHNS, E. R. & BEAVEN, G. H. (1971). Developmental changes in human haemoglobins. In *The Biochemistry of Development*, ed. BENSON, P., pp. 175–203. London: Spastics International Medical Publications and London: Heinemann Medical.

HUGGETT, A. ST G. (1927). Foetal blood gas tensions and gas transfusion through the placenta of the goat. *J. Physiol.* **62**, 373–384.

HUGGETT, A. ST G. (1959). Comparative foetal blood sugars. *J. Physiol.* **146**, 53–54P.

HULL, D. (1974). The function and development of adipose tissue. In *The Scientific Foundations of Paediatrics*, ed. DAVIS, J. A. & DOBBING, J., pp. 440–455. London: Heinemann Medical.

HULL, D. (1975). Storage and supply of fatty acids before and after birth. *Br. med. Bull.* **31**, 32–36.

HUNGERLAND, H. (1958). *Arzt und Wissenschaft*. Inaugeral Lecture, Freidrich-Wilhelms-Universität, Bonn.

JACKSON, C. M. (1928). Some aspects of form and growth. In *Growth*, by Robbins, W. J., Brody, S., Hogan, A. G., Jackson, C. M. & Greene, C. W., pp. 111–140. New Haven: Yale University Press and London: Humphrey Milford, Oxford University Press.

JOLLIFFE, N., SHANNON, J. A. & SMITH, H. W. (1932). The excretion of urine in the dog. 3. The use of non-metabolized sugars in the measurement of the glomerular filtrate. *Am. J. Physiol.* **100**, 301–312.

162   R. A. McCANCE

JONXIS, J. H. P., VISSER, H. K. A. & TROELSTRA, J. A. (ed.) (1964). *The Adaptation of the Newborn Infant to Extra-uterine Life*, Nutricia Symposium. Leiden: Stenfert Kroese.

JOPPICH, G. & WOLF, H. (1958). Reststickstoffer-Höhungen in Blut von Frühgeborenen in den ersten Lebenstagen. *Klin. Wschr.* **36**, 616–619.

KENNEDY, G. C. (1957). The development with age of hypothalamic restraint upon the appetite of the rat. *J. Endocrinol.* **16**, 9–17.

KERPEL-FRONIUS, E. (1940). Theoretische und praktishe Bemerkungen zur Flüssigheitsbehandlung von Exsiccationszuständen *Mschr. Kinderheilkd.* **81**, 294–304.

KERPEL-FRONIUS, E. (1959). *Pathologie und Klinik des Salz und Wasserhaushalts.* Budapest: Ungarischen Akademie des Wissenschaften.

KLING, A. (1964). Effects of rhinencephalic lesions on endocrine and somatic development in the rat. *Am. J. Physiol.* **206**, 1395–1400.

LADIMER, I. (1955). Ethical and legal aspects of medical research on human beings. *J. Public Law* **3**, 467–511.

LADIMER, I. & NEWMAN, R. W. (1963). *Clinical Investigation in Medicine: Legal, Ethical and Moral aspects.* Boston: Law–Medicine Research Institute, Boston University.

LANMAN, J. T. (ed.) (1956–1960). *Conferences on Physiology of Prematurity.* Josiah Macy Foundation. Madison: Madison Printing Co.

LAWN, L. & McCANCE, R. A. (1962). Ventures with an artificial placenta 1. Principles and preliminary results. *Proc. R. Soc.* B **155**, 500–509.

LAWN, L., McCANCE, R. A. & THORN, A. E. (1967). Artificial placentae: Comparative results with two gas exchangers. *Quart. J. exp. Physiol.* **52**, 157–167.

LIGGINS, G. C. (1973). Foetal participation in the physiological controlling mechanisms of parturition. In *Barcroft Centenary Symposium*, ed. COMLINE, R. S., CROSS, K. W., DAWES, G. S. & NATHANIELSZ, P. W., pp. 298–305. London: Cambridge University Press.

LISTER, D., COWEN, T. & McCANCE, R. A. (1966). Severe undernutrition in growing and adult animals. 16. The ultimate results of rehabilitation: poultry. *Br. J. Nutr.* **20**, 663–639.

LISTER, D. & McCANCE, R. A. (1967). Severe undernutrition in growing and adult animals. 17. The ultimate results of rehabilitation: pigs. *Br. J. Nutr.* **21**, 787–798.

LIU, S. H., CHU, H. I., HSU, H. C., CHAO, H. C. & CHEU, S. H. (1941). Calcium metabolism in osteomalacia. x. The pathogenic role of pregnancy and the importance of calcium and vitamin D supply. *J. clin. Invest.* **20**, 255–271.

LOHER, W. (1960–1961). The chemical acceleration of the maturation process and its hormonal control in the male of the desert locust. *Proc. R. Soc.* B **153**, 380–397.

LUCEY, J., FERRIERO, M. & HEWITT, J. (1968). Prevention of hyperbilirubinemia of prematurity by phototherapy. *Pediatrics* **41**, 1047–1054.

McCANCE, R. A. (1935). The effect of sudden severe anoxaemia on the function of the human kidney. *Lancet* **ii**, 370–372.

McCANCE, R. A. (1936). Experimental sodium chloride deficiency in man. *Proc. R. Soc.* B **119**, 245–268.

McCANCE, R. A. (1948). Renal function in early life. *Physiol. Rev.* **28**, 331–348.

McCANCE, R. A. (1950). Renal physiology in infancy. *Am. J. Med.* **9**, 229–241.

McCANCE, R. A. (1951). The practice of experimental medicine. *Proc. R. Soc. Med.* **44**, 189–194.

McCANCE, R. A. (1959a). *Reflections of a Medical Investigator.* Scripta Academica Groningana. Groningen: J. B. Wolters.

McCANCE, R. A. (1959b). The maintenance of stability in the newly born: 1. Chemical exchange 2. Thermal balance. *Arch. Dis. Childh.* **34**, 361–370, 459–470.

McCance, R. A. (1960). The development of acid–base control. In *The Development of Homeostasis with Special Reference to Factors of the Environment*, pp. 49–54. Prague: Czechoslovak Academy of Sciences.

McCance, R. A. (1968). The effect of calorie deficiencies and protein deficiencies on final weight and stature. See *Calorie Deficiencies and Protein Deficiencies*, ed. McCance R. A. & Widdowson, E. M., pp. 319–328. London: Churchill.

McCance, R. A. (1972). The role of the developing kidney in the maintenance of internal stability. *J. R. Coll. Phys.* **6**, 235–245.

McCance, R. A. & Hatemi, N. (1961). Control of acid–base stability in the newly born. *Lancet* i, 293–297.

McCance, R. A. & Madders, K. (1930). The comparative rates of absorption of sugars from the human intestine. *Biochem. J.* **24**, 795–804.

McCance, R. A., Naylor, N. J. B. & Widdowson, E. M. (1954). The response of infants to a large dose of water. *Arch. Dis. Childh.* **29**, 104–109.

McCance, R. A. & Otley, M. (1951). Course of the blood urea in newborn rats, pigs and kittens. *J. Physiol.* **113**, 18–22.

McCance, R. A. & Strangeways, W. M. B. (1954). Protein katabolism and oxygen consumption during starvation in infants, adults and old men. *Br. J. Nutr.* **8**, 21–32.

McCance, R. A. & Widdowson, E. M. (1936). The response of the kidney to an alkalosis during salt deficiency. *Proc. R. Soc.* B **120**, 228–239.

McCance, R. A. & Widdowson, E. M. (1939). The fate of strontium after intravenous administration to normal persons. *Biochem. J.* **33**, 1822–1825.

McCance, R. A. & Widdowson, E. M. (1947). Blood urea in the first nine days of life. *Lancet* i, 787–788.

McCance, R. A. & Widdowson, E. M. (1954). Water metabolism. *Cold Spring Harbor Symp. Quant. Biol.* **19**, 155–160.

McCance, R. A. & Widdowson, E. M. (1956). Metabolism, growth and renal function of piglets in the first days of life. *J. Physiol.* **133**, 373–384.

McCance, R. A. & Widdowson, E. M. (1957). The effect of food and growth on the stability of the internal environment. *Mod. Prob. Pediatrics* **2**, 179–186.

McCance, R. A. & Widdowson, E. M. (1958a). Hypertonic expansion of the extracellular fluids. *Acta Paediat.* **46**, 337–353.

McCance, R. A. & Widdowson, E. M. (1958b). The response of the newborn puppy to water, salt and food. *J. Physiol.* **141**, 81–87.

McCance, R. A. & Widdowson, E. M. (1959). The effect of lowering the ambient temperature on the metabolism of the newborn pig. *J. Physiol.* **147**, 124–134.

McCance, R. A. & Widdowson, E. M. (1974). The determinants of growth and form. *Proc. R. Soc.* B **185**, 1–17.

McCance, R. A. & Wilkinson, E. (1947). The response of adult and suckling rats to the administration of water and of hypertonic solutions of urea and salt. *J. Physiol.* **106**, 256–263.

McCance, R. A. & Young, W. F. (1941). The secretion of urine by newborn infants. *J. Physiol.* **99**, 265–282.

McCarthy, E. F. (1934). A comparison of foetal and maternal haemoglobins in the goat. *J. Physiol.* **80**, 206–212.

McCay, C. M., Maynard, L. A., Sperling, G. & Barnes, L. L. (1939). Retarded growth, life span, ultimate body size and age changes in the albino rat after feeding diets restricted only from weaning. *J. Nutr.* **18**, 1–14.

McClelland, R. J. & Cowley, J. J. (1972). The effects of lesions of the olfactory bulbs on the growth and behaviour of mice. *Physiol. Behav.* **9**, 319–324.

MacCONNACHIE, H. F., ENESCO, M. & LEBLOND, C. P. (1964). The mode of increase in the number of skeletal muscle nuclei in the postnatal rat. *Am. J. Anat.* **114**, 245–253.

McLAREN, A. (1965). Genetic and environmental effects on foetal and placental growth in mice. *J. Reprod. Fertil.* **9**, 79–98.

McLAREN, A. & MICHIE, D. (1960). Control of pre-natal growth in mammals. *Nature, Lond.* **187**, 363–365.

MAKEPIECE, A. W., FREMONT-SMITH, F., DAILEY, M. E. & CARROLL, M. P. (1931). Nature of amniotic fluid: comparative study of human amniotic fluid and maternal serum. *Surgery Gynec. Obstet.* **53**, 635–644.

MATTHEWS, B. *et al.* (1954–1955). A discussion on the physiology of man at high altitudes. *Proc. R. Soc.* B **143**, 1–42.

MAXWELL, J. P. (1930a). Further studies in osteomalacia. *Proc. R. Soc. Med.* **23**, 639–652.

MAXWELL, J. P. (1930b). Two cases of foetal rickets. *J. Path. Bact.* **33**, 327–338.

MESCHIA, G., WOLKOFF, A. S. & BARRON, D. H. (1959). The oxygen carbon dioxide and hydrogen-ion concentrations in the arterial and uterine venous bloods of pregnant sheep. *Quart. J. exp. Physiol.* **44**, 333–342.

MESCHIA, G., COTTER, J. R., BREATHNACH, C. S. & BARRON, D. H. (1965). The hemoglobin, oxygen, carbon dioxide and hydrogen ion concentrations in the umbilical bloods of sheep and goats as sampled via indwelling plastic catheters. *Quart. J. exp. Physiol.* **50**, 185–195.

MESTYAN, J., JARAI, F., BATA, G. & FEKELE, M. (1964). Surface temperature versus deep body temperature and the metabolic response to cold of hypothermic premature infants. *Biol. Neonat.* **7**, 243–254.

MILLER, J. F. A. P. (1966). Immunity in the foetus and the newborn. *Br. med. Bull.* **22**, 21–26.

MOLESCHOTT, J. (1859). *Physiologie der Nahrungsmittel: ein Handbuch der Diatetik.* Giessen: Ferbersche Universitatsbuchhandlung.

MÖLLER, E., McINTOSH, J. F. & van SLYKE, D. D. (1929). Studies on urea excretion: 2. Relationship between urine volume and the rate of urea excretion by normal adults. *J. clin. Invest.* **6**, 427–465.

MOLLISON, P. L. & CUTBUSH, M. (1949). Bromsulphthein excretion in the newborn. *Arch. Dis. Childh.* **24**, 7–11.

MOORE, R. E. & UNDERWOOD, M. C. (1963). The thermogenic effects of noradrenaline in newborn and infant kittens and other small mammals. A possible hormonal mechanism in the control of heat production. *J. Physiol.* **168**, 290–317.

MORRIS, B. (1975). The transmissions of ¹²⁵I-labelled IgG by proximal and distal regions of the small intestine of 16 day old rats. *J. Physiol.* **245**, 249–259.

MORRIS, I. G. (1965–1966). The transmission of anti-*Brucella abortus* agglutinins across the gut in young rats. *Proc. R. Soc.* B **163**, 402–416.

MOUNT, L. E. (1963). Responses to thermal environment in newborn pigs. *Fedn Proc.* **22**, 818–823.

MOUNT, L. E. (1964). Tissue and air components of thermal insulation in the newborn pig. *J. Physiol.* **170**, 286–295.

NADAL, J. W., PEDERSEN, S. & MADDOCK, W. G. (1941). A comparison between dehydration from salt loss and from water deprivation. *J. clin. Invest.* **20**, 691–703.

NEEDHAM, J. (1931). *Chemical Embryology*, vols. **1**, **2**, and **3**. Cambridge: Cambridge University Press.

NIXON, D. A., BRITTON, H. G. & ALEXANDER, D. P. (1963). Perfusion of the viable sheep foetus. *Nature, Lond.* **199**, 183–185.

O'SHAUGHNESSY, W. B. (1832). Report on the Chemical Pathology of the Malignant Cholera. London: Highley. (Published by authority of the Central Board of Health.) See Tracts B 107 1–118, as bound at the Royal Society of Medicine.

PAPPWORTH, M. H. (1967). Human Guinea-pigs: Experimentation on Man. London: Routledge & Kegan Paul. (Reprinted, 1969, Harmondsworth: Penguin.)

PARSONS, L. (1950). The Influence of Harvey and his Contemporaries on Paediatrics. Harveian oration to the Royal College of Physicians. London: Headley Brothers.

PATTLE, R. E. (1958). Properties, function and origin of the alveolar lining layer. Proc. R. Soc. B 148, 217–240.

PATTLE, R. E. (1965). Surface lining of lung alveoli. Physiol. Rev. 45, 48–79.

PAVEY, D. E. & WIDDOWSON, E. M. (1975). Influence of dietary fat intake of the mothers on the composition of body fat of newborn guinea-pigs. Proc. Nutr. Soc. 34, 107A–108A.

PERRY, J. S. & STANIER, M. W. (1962). The rate of flow of urine of foetal pigs. J. Physiol. 161, 344–350.

PFLÜGER, E. (1868). Ueber die Ursache der Athembewegungen sowie der Dyspnoë und Apnoë. Pflügers Arch. ges Physiol. 1, 61–106.

PFLÜGER, E. (1877). Die Lebenszähigkeit des menschlichen Foetus. Pflügers Arch. ges. Physiol. 14, 628–629.

PIERCE, A. E. & SMITH, M. W. (1967). The in vitro transfer of immune lactoglobulin of newborn pigs. J. Physiol. 190, 19–34.

POLLAK, O. (1869). Beitrage zur Kentnis des Harnes der Säuglinge. Jahrb. Kinderheilkd. N. F. 2, 27–32.

PONTE, J. & PURVES, M. J. (1973). Types of afferent nervous activity which may be measured in the vagus nerve of the sheep foetus. J. Physiol. 229, 51–76.

PORTER, A. M. W. (1973). Minors and medical experiments. Br. med. J. 1, 46–47.

PRECHTL, H. F. R. & LENARD, H. G. (1967). A study of eye movements in sleeping newborn infants. Brain Res. 5, 477–493.

PRIESTLEY, J. G. (1921). The regulation of the excretion of water by the kidneys. J. Physiol. 55, 305–318.

PROUT, W. (1831). Observations on the application of chemistry to physiology, pathology and practice. Lond. med. Gaz. 8, 257–265.

REHBERG, P. B. (1926a). Studies on kidney function: 1. The rate of filtration and reabsorption in the human kidney. Biochem. J. 20, 447–460.

REHBERG, P. B. (1926b). Studies on kidney function: 2. The excretion of urea and chlorine analysed according to a modified filtration–reabsorption theory. Biochem. J. 20, 461–482.

REYNOLDS, E. O. R. (1975). The management of hyaline membrane disease. Br. med. Bull. 31, 18–23.

RHODES, P. (1975a). Letter from South Australia: afflictions and admonitions. Br. med. J. 4, 340–341.

RHODES, P. (1975b). Inflation down under. Br. med. J. 4, 566–567.

ROGERS, L. (1909). A second season's experience of hypertonic transfusions in cholera controlled by observations on the blood changes. Therap. Gaz. 33 (Ser 3, 25), 761–767.

ROSEN, G. D. & FERNELL, W. R. (1956). Microbiological evaluation of protein quality with Tetrahymena pyriformis W. 2. Relative nutritive values of proteins in foodstuffs. Br. J. Nutr. 10, 156–168.

RUBIN, B. L., CALCAGNO, P. L., RUBIN, M. I. & WEINTRAUB, M. D. (1956). Renal defence response to induced acidosis in premature infants (ammonia production and titratable acid excretion). Am. J. Dis. Child. 92, 513.

RUBIN, M. I., BRUCK, E. & RAPOPORT, M. (1949). Maturation of renal function in childhood: clearance studies. *J. clin. Invest.* **28**, 1144–1162.

SABRAZES & FAUQUET. (1901). Propriétés hématolytique de la première urine du nouveau-né. *C. r. Séanc. Soc. Biol.* **12**, 372.

SCAMMON, R. E. (1927). The first seriatim study of human growth. *Am. J. phys. Anthropol.* **10**, 329–336.

SCHNIEDEN, H. (1957). Water diuresis in children aged 1–3 years. *Arch. Dis. Childh.* **32**, 189–912.

SHANNON, J. A. (1934). The excretion of inulin by the dogfish, *Squalus acanthias*. *J. cell. comp. Physiol.* **5**, 301–310.

SHANNON, J. A. (1935). The excretion of inulin by the dog. *Am. J. Physiol.* **112**, 405–413.

SHANNON, J. A., JOLLIFFE, N. & SMITH, H. W. (1932). The excretion of urine in the dog. 6. The filtration and secretion of exogenous creatinine. *Am. J. Physiol.* **102**, 534–550.

SHANNON, J. A. & SMITH, H. W. (1935). The excretion of inulin, xylose and urea by normal and phlorizinized man. *J. clin. Invest.* **14**, 393–401.

SHAW, J. C. L. (1973). Special problems of feeding very low birth weight infants. See *Nutritional Problems in a Changing World*, ed. HOLLINGSWORTH, D. & RUSSELL, M., pp. 115–124. Barking, Essex: Applied Science Publishers.

SHAW, J. C. L. (1974). Malnutrition in very low birth weight, pre-term infants. *Proc. Nutr. Soc.* **33**, 103–112.

SHELLEY, H. J. (1960). Blood sugars and tissue carbohydrate in foetal and infant lambs and rhesus monkeys. *J. Physiol.* **153**, 527–552.

SHELLEY, H. J. (1962). Glycogen reserves and their changes at birth and in anoxia. *Br. med. Bull.* **17**, 137–143.

SHELLEY, H. J. (1964). Carbohydrate reserves in the newborn infant. *Br. med. J.* **1**, 273–275.

SHELLEY, H. J., BASSETT, J. M. & MILNER, R. D. G. (1975). Control of carbohydrate metabolism in the fetus and newborn. *Br. med. Bull.* **31**, 37–43.

SHELLEY, H. J. & NELIGAN, G. A. (1966). Neonatal hypoglycaemia. *Br. med. Bull.* **22**, 34–39.

SILVERMAN, W. A., AGATE, F. J. & FERTIG, J. W. (1963). A sequential trial of the non-thermal effect of atmospheric humidity on survival of newborn infants of low birth weight. *Pediatrics* **31**, 719–724.

SILVERMAN, W. A., FERTIG, J. W. & BERGER, A. P. (1958). The influence of the thermal environment upon the survival of newly born premature infants. *Pediatrics* **22**, 876–886.

SMITH, C. A. (1945, 1959). *The Physiology of the Newborn Infant*, 1st edn (1945), 3rd edn (1959). Springfield: Charles Thomas.

SMITH, C. A. (1947). Effects of wartime starvation in Holland on pregnancy and its products. *Am. J. Obstet. Gynec.* **53**, 599–608.

SMITH, R. E. & HOIJER, D. J. (1962). Metabolism and cellular function in cold acclimation. *Physiol. Rev.* **42**, 60–142.

SMITH, R. E. & HOCK, R. J. (1963). Brown fat: thermogenic effector of arousal in hibernators. *Science*, **140**, 199–200.

SMITH, R. E. & HORWITZ, B. A. (1969). Brown fat and thermogenesis. *Physiol. Rev.* **49**, 330–425.

SMITH, R. E. & ROBERTS, J. C. (1964). Thermogenesis of brown adipose tissue in cold-acclimated rats. *Am. J. Physiol.* **206**, 143–148.

SMITH, R. T. (1968). Development of fetal and neonatal immunological function. In *The Biology of Gestation*, vol. **2**, ed. ASSALI, N.S., pp. 321–354. New York and London: Academic Press.

SNELLING, C. E. (1943). Disturbed kidney function in the newborn infant associated with a decreased calcium/phosphorus ratio. *J. Pediat.* **22**, 550–564.

STANIER, M. (1971). Osmolarity of urine from renal pelvis and bladder of foetal and post-natal pigs. *J. Physiol.* **218**, 30–31P.

STEIN, Z., SUSSER, M., SAENGER, G. & MAROLLA, F. (1975). *Famine and Human Development: The Dutch Hunger Winter of 1944–1945.* New York and London: Oxford University Press.

DE SWIET, M., FANCOURT, R. & PETO, J. (1975). Systolic blood pressure variation during the first six days of life. *Clin. Sci. Molec. Med.* **49**, 557–561.

TANNER, J. M. (1962). *Growth at Adolescence,* 2nd edn. Oxford and Edinburgh: Blackwell.

TAUSCH, M. (1936). Der Fetalharn. *Arch. Gynaek.* **162**, 217–267.

THOMSON, A. M. (1968). The later results in man of malnutrition in early life. In *Calorie Deficiencies and Protein Deficiencies,* ed. McCANCE, R. A. & WIDDOWSON, E. M., pp. 289–299. London: Churchill Livingstone.

THOMSON, J. (1944). Observations on the urine of the newborn infant. *Arch. Dis. Childh.* **19**, 169–177.

TIMIRAS, P. S. (1972). *Developmental Physiology and Aging.* New York: Macmillan.

TONGE, H. C. & McCANCE, R. A. (1973). Normal development of the jaws and teeth in pigs and the delay and malocclusion produced by calorie deficiencies. *J. Anat.* **115**, 1–22.

VALMAN, H. B., HEATH, C. D. & BROWN, R. K. J. (1972). Continuous intragastric milk feeds in infants of low birth weight. *Br. med. J.* **3**, 547–550.

VARGA, F. (1959). The respective effects of starvation and changed body composition on energy metabolism in malnourished infants. *Pediatrics* **23**, 1085–1190.

VESTERDAL, J. & TUDVAD, F. (1949). Studies on the kidney function in premature and full term infants by estimation of the inulin and para-amino-hippurate clearances. *Acta Paediat.* **37**, 429–440.

VIERORT, H. (1880). *Anatomische, physiologische und physikalische Daten und Tabellen zum Gebrauche für Mediciner.* Jena: Gustav Fischer.

VOISIN, G. A. & CHAOVAT, G. (1974). Demonstration, nature and properties of maternal antibodies fixed on placenta and directed against paternal antigens. *J. Reprod. Fertil. Suppl.* **21**, 89–103.

WALKER, H. M. & DANESH, J. N. Z. (1973). Extracorporeal circulation for the study of the full term fetus. In *The Mammalian Foetus In Vitro,* ed. AUSTIN, C. R., pp. 209–250. London: Chapman & Hall.

WALTON, A. & HAMMOND, J. (1938). The maternal effects on growth and conformation in Shire horse–Shetland pony crosses. *Proc. R. Soc.* B **125**, 311–334.

WEST, J. R., SMITH, H. W. & CHASIS, H. (1948). Glomerular filtration rate, effective renal blood flow, and maximum tubular excretory capacity in infancy. *J. Pediat.* **32**, 10–18.

WESTIN, B., NYBERG, R. & ENHÖRNING, G. (1958). A technique for perfusion of the previable human fetus. *Acta Paediat.* **47**, 339–349.

WHALAN, D. J. (1975). The ethics and morality of clinical trials in man. *Med. J. Australia* **1**, 491–494.

WIDDOWSON, E. M. (1965). Chemical analysis of the body. In *Human Body Composition,* ed. BROZEK, J., pp. 31–47. New York and London: Pergamon.

WIDDOWSON, E. M. (1968). Growth and composition of the fetus and newborn. In *Biology of Gestation,* vol. 2, ed. ASSALI, N. S., pp. 1–49. New York and London: Academic Press.

WIDDOWSON, E. M. (1971). Intra-uterine growth retardation in the pig. 1. Organ size and cellular development at birth and after growing to maturity. *Biol. Neonate* **19**, 329–340.

WIDDOWSON, E. M., CRABB, D. E. & MILNER, R. D. G. (1972). Cellular development of some human organs before birth. *Arch. Dis. Childh.* **47**, 652–655.

WIDDOWSON, E. M., DAUNCEY, M. J., GAIRDNER, D. M. T., JONXIS, J. H. P. & PELIKAN-FILIPKOVA, M. (1975). Body fat of British and Dutch infants. *Br. med. J.* **1**, 653–655.

WIDDOWSON, E. M. & DICKERSON, J. W. T. (1964). The chemical composition of the body. In *Mineral Metabolism*, ed. COMAR, C. L. & BRONNER, F., vol. 2A, pp. 1–247. New York and London: Academic Press.

WIDDOWSON, E. M. & McCANCE, R. A. (1960). Some effects of accelerating growth. 1. General somatic development. *Proc. R. Soc.* B **152**, 188–206.

WIDDOWSON, E. M. & McCANCE, R. A. (1963). The effects of finite periods of under-nutrition at different ages on the composition and subsequent development of the rat. *Proc. R. Soc.* B **158**, 329–342.

WIDDOWSON, E. M. & McCANCE, R. A. (1975). A review: new thoughts on growth. *Pediat. Res.* **9**, 154–159.

WILKINSON, A. W. (1973). Some aspects of renal function in the newly born. *J. Pediat. Surg.* **8**, 103–116.

WILLIAMS, R. B. & BREMNER, I. (1976). Copper and zinc deposition in the foetal lamb. *Proc. Nutr. Soc.* **35**, 86A–88A.

WILLIAMSON, R. C. & HIATT, E. P. (1947). Development of renal function in fetal and neonatal rabbits using phenolsulphonthalein. *Proc. Soc. exp. Biol. Med.* **66**, 554–557.

WINICK, M. (1972). *Nutrition and Development*, ed. WINICK, M., pp. 49–98. New York and London: Wiley.

WINICK, M., BRASEL, J. A. & ROSSO, P. (1972). Nutrition and cell growth. In *Nutrition and Development*, ed. WINICK, M., pp. 49–98. New York and London: Wiley.

WINICK, M. & NOBLE, A. (1965). Quantitative changes in DNA, RNA and protein during prenatal and postnatal growth in the rat. *Devel. Biol.* **12**, 451–466.

WINICK, M. & NOBLE, A. (1966). Cellular response in rats during malnutrition at various ages. *J. Nutr.* **89**, 300–306.

WLADIMIROFF, J. W. & CAMPBELL, S. (1974). Fetal urine-production rates in normal and complicated pregnancy. *Lancet* **i**, 151–154.

WOLSTENHOLME, G. E. S. & O'CONNOR, M. (ed.) (1961). *Somatic Stability of the Newly Born*, CIBA Foundation Symposium. London: Churchill.

YOUNG, W. F., HALLUM, J. L. & McCANCE, R. A. (1941). The secretion of urine by premature infants. *Arch. Dis. Childh.* **16**, 243–252.

YUDKIN, S., GELLIS, S. S. & LAPPEN, F. (1949). Liver function in newborn infants with special reference to excretion of bromsulphalein. *Arch. Dis. Childh.* **24**, 12–14.

ZETTERSTRÖM, R. (1961). Carbohydrate metabolism and the role of the liver. In *Somatic Stability in the Newly Born*, CIBA Foundation Symposium, ed. WOLSTENHOLME, G. E. W. & O'CONNOR, M., pp. 59–74. London: Churchill.

ZUNTZ, N. (1877). Ueber die Respiration des Saugethier-Foetus. *Pflügers Arch. ges. Physiol.* **14**, 605–627.

ZWEYMULLER, E., WIDDOWSON, E. M. & McCANCE, R. A. (1959). The passage of urea and creatinine across the placenta of the pig. *J. Embryol. exp. Morphol.* **7**, 202–209.

NAMES INDEX

SUBJECT INDEX

acetylcholine: contractures produced in denervated striated muscle by, 66; muscarine-like and nicotine-like effects of, 65; opposition to theory of transmission by, 67, 72; paralysis caused by excess of, 72, 80; synthesis of, 72, 80; as transmitter of nerve impulse, in central nervous system, 80, in electric organ of fishes, 73–5; in ganglia, 71, 72, 75–6, 78, 79–80, to muscle, 71, 72, 77, 78, 80–2, and in parasympathetic and sympathetic systems, 75–6, 77, 78, 83

actin: in muscle, 31, 54; in other types of cell, 41

Actinosphaerium, birefringence in contractile threads of, 52

action potential in nerve, 7; compared with resting potential, 11, 12–13, 14; computation of, 19; effect of ions on, 14, 16; measurement of, with internal electrode, 10, 12, 13; theoretical system giving rise to, 15, 18

adenylate cyclase, in hormone action, 126

adrenal glands: acetylcholine in response of, to splanchnic stimulation, 75, 76, 77, 83

adrenergic, introduction of term, 76–7

amoebic movement, 37, 40

apparatus, construction of, 5–6, 10–11, 17

arthropods, varying width of muscle striations in, 34

atropine: blocks action of vagus on pancreas, 107, but not on stomach and bladder, 75; prevents action of acetylcholine (not its release) in autonomic nervous system, 77, 79; without effect on response of pancreas to hydrochloric acid in duodenum, 107, or to cholecystokinin, 117

axon, introduction of electrode into, 12–13

axoplasm: liquefied by calcium, 18; normally solid, 12

bilirubin: destruction of, by light, 153; maturation of system for conversion of, to glucuronide, 152

birefringence: association of 'contractility' with, 37, 50–3; increases with stretching, except in muscle, 50; in muscle during contraction and elongation, 26, 34, 52; present in muscle only in A bands, 50; quantitative analysis of, 49n

brain: growth of, 147; receptive fields on retina, and nerve cells in cortex of, 96–7

bridges: early success and later accidents for each type of, 58

calcium: and contraction of spasmoneme of *Vorticella*, 41–2; liquefaction of axoplasm by, 18; in theory of nerve activity, 15

cells: division and enlargement of, during growth, 147–8

central nervous system: acetylcholine transmission in, 80; see also brain

chemical transmission of nerve impulse, slow acceptance of theory of, 66–7, 72

cholecystokinin (pancreozymin), hormone induced by fat in duodenum, 116–17; demonstrated in blood by radioimmunoassay, 122; effects of, 118–19; potentiation of secretin by, 120; structure of, 118, (longer variant) 122

cholesterol, in serum of Dutch and British babies, 152

choline, and acetylcholine synthesis, 80

cholinergic, introduction of term, 76–7

cholinesterase: in electric organ of fishes, 73; prediction and discovery of presence of, in blood, 65; see also eserine, inhibitor of cholinesterase

ciné micrography, of muscle fibres, 44, 48

'clap and fling' method of generating lift, used by very small insects, 41n

colour vision, 89, 100–1; colour-opponent processes in, 104; in dichromats, 101–2; two-colour projections and, 103–4; two-colour thresholds and, 101

conductance curves, for nerve, 10, 19

contractile mechanisms: birefringence in, 37, 50–3; theory of essential uniformity of, 37–43, 55

contraction bands, in muscle fibres, 30–1, 32, 34, 45

copper, in foetal liver at term, 146

creatinine, impermeability of placenta to, 140

curarine: prevents action of acetylcholine (but not its release) in motor nerve stimulation, 79; renders ganglionic cells insensitive to acetylcholine but not to potassium, 80

deuteranopes, lack green-sensitive cone pigment, 102

dichromats (red-green colour-blind persons), 102–3

diet: of new-born mammals, and renal function, 140–1

digestive system: as complex endocrine gland, 128; Pavlov and, 105–6

diuresis: after administration of water, or of sodium chloride or urea solutions, in adult and new-born rats, 138

DNA: total amount of, and ratio of, to protein, in muscle during growth, 148

duodenum: effects of fat in, 115–16; effects of hydrochloric acid in, 106, 123; extract from (incretin), and insulin secretion by pancreas, 114–15, 119, 120

electric organ of fishes, cholinergic nature of nerves to, 68, 73–4

electric recordings: from optic nerve, 94; from single-cell photoreceptors, 98–100

electroencephalography, 93–4

electron microscopy: of muscle structure, 39n, 49–50, 53, 57; osmium tetroxide as fixative for, 47–50

electroretinogram, measure of retinal organization, 94–6

endocrine glands: development of, in foetus, 144–5

'enterocrinin', 116, 119

'enterogastrone', 115–16, 119

enteroglucagon, candidate hormone, 124–5

eserine: inhibition of cholinesterase by, 65, 69, 71–2, 75, 82, 83; leech muscle treated with, for detection of acetylcholine, 69, 71

ethics of experiments on humans, 141, 155; Claude Bernard on, 134, 136

eye: dark adaptation of, 89–90; in vertebrates and invertebrates, 90; see also retina

fat: brown, in maintenance of temperature in new-born animals, 143, 151; percentage

of, in body, at different ages, 145–6, and in new-born animals, 151

fatty acids: passage of, through placenta, 151–2

fishes, electric organ of, 68, 73–4

fixatives, and cell structure as seen by microscope, 46, 47, 48

flagella, 41; birefringence of, 51

foetus: accumulation of glycogen in, 150–1; development of endocrine glands in, 144–5; haemoglobin of, 137; immature, possibility of keeping alive and functional, 143–4; rickets in, 135; survival of anoxia by, 136, 142; as 'tolerated allograft', 154; urine production by, 140

fovea (blue-blind), red- and green-sensitive cones in, 89

fructose, in fluids of foetal calf, 133–4

gall bladder: contraction of, caused by cholecystokinin, 116, 117, and by gastrin, 118

ganglia of autonomic nervous system: acetylcholine in transmission in, 71, 72, 75–6, 78, 79–80; perfusion of, 76, 83

gastric inhibitory polypeptide, from duodenum, 117–18; demonstrated in blood by radioimmunoassay, 122; effects of, 119, 120

gastrin, gastric hormone, 110–11; effects of, 118, 120; isolation and structure of, 112; radioimmunoassay of two forms of, 121–2

glycogen, accumulates in foetus, 150–1

growth: cell division and cell enlargement in, 147–8; change in composition of body during, 145–7; curve of, in different animals, 145; undernutrition and, 148–50

haemoglobin, foetal, 137

histamine, stimulation of gastric acid secretion by, 111–12

Holland, C_{18} fatty acids in body fat of infants in, 152

hormones: multiple actions of, 118–19; origin of term, 110; photoaffinity labelling of binding sites for, 127–8; radioimmunoassay of, in blood, 119, 120, 121–2; radiolabelling of, 126; receptors for, 125–8

humans, experiments on, 134–5, 136; ethics of, 134, 136, 141, 155

hydrochloric acid in duodenum, induces secretion of pancreatic juice, 106, 123

hypercalcaemia in infants, caused by excess of vitamin D, 135

hypoglycaemia, better tolerated by new-born than by adult animals, 151

immune bodies: passage of, from mother to young, 154

Uniformity of Nature, effects on muscle theory of supposed principle of, 37–43, 55

urea, in plasma of new-born humans, pigs, and rats, 139–40

vagus nerve, and secretions of digestive tract, 106, 120

vasoactive intestinal peptide, candidate hormone, 124

Vernier–Morrison syndrome, vasoactive intestinal peptide in, 124

vertebrates: response to light of eyes of, 99; width of muscle striations in, 34; vesicles, in muscle, 39n

'villikinin', 116

vision, photochemical theory of, 90–1; *see also* colour vision

visual pigments, 80

vitamin D, excess of, 135

vitamins, 135

vorticellids, spasmoneme (contractile thread) of, 41; birefringence of, 51; calcium and, 41–2

water: percentage of, in body at different ages, 145

Zoothamnion, spasmoneme of, 41, 52